MANGROVES AND SEDIMENT DYNAMICS ALONG THE COASTS OF SOUTHERN THAILAND

Promotor: Prof. dr. Patrick Denny
                       Professor of Aquatic ecology
                       UNESCO-IHE /Wageningen University, The Netherlands

Co-promotor: Dr. ir. J. E. Vermaat
                       Associate Professor Aquatic and Wetland Ecology
                       Free University Amsterdam,
                       The Netherlands

Awarding Committee: Prof. dr. N. Koedam
                       Free University Brussels, Belgium

                       Prof. dr. M. Scheffer
                       Wageningen University, The Netherlands

                       Prof. dr. J. O'Keeffe
                       UNESCO-IHE, Delft, The Netherlands

                       Prof. dr. W.J. Wolff
                       University of Groningen, The Netherlands

MANGROVES AND SEDIMENT DYNAMICS ALONG THE COASTS OF SOUTHERN THAILAND

Udomluck THAMPANYA
born in Chiang Mai, Thailand

DISSERTATION

Submitted in fulfillment of the requirements of
the Academic Board of Wageningen University and
the Academic Board of the UNESCO-IHE Institute for Water Education
for the Degree of DOCTOR
to be defended in public
on Monday, 27 March 2006 at 15:00 h in Delft, The Netherlands

CRC Press
Taylor & Francis Group
Boca Raton London New York

CRC Press is an imprint of the
Taylor & Francis Group, an **informa** business

CRC Press
Taylor & Francis Group
6000 Broken Sound Parkway NW, Suite 300
Boca Raton, FL 33487-2742

First issued in hardback 2017

© 2006 by Taylor & Francis Group, LLC
CRC Press is an imprint of Taylor & Francis Group, an Informa business

No claim to original U.S. Government works

ISBN-13: 978-0-415-3846-5 (pbk)
ISBN-13: 978-1-138-43392-2 (hbk)

Transferred to Digital Printing 2007

**Visit the Taylor & Francis Web site at
http://www.taylorandfrancis.com**

**and the CRC Press Web site at
http://www.crcpress.com**

Preface

This PhD research project originated from the EC-funded CERDS project (Coastal Ecosystems Response to Deforestation-derived Siltation in Southeast Asia), which investigated the impacts of siltation on several coastal ecosystem components. It was a joint project between researchers from various institutions in Europe (UNESCO IHE-Netherlands, CSIC-Spain, FBL-Denmark) and Asia (CORIN-Thailand and MSI-Philippines). Having participated mostly in the mangrove component, I became fascinated by these special plants, not only their peculiar characteristics but also their important values and functions for our planet. After the CERDS project ended in 1997, my PhD research on mangroves was made possible with the help of Jan Vermaat in WOTRO-fellowship acquisition.

I should like to take this opportunity to express my gratitude to the Netherlands Foundation for the Advancement of Tropical Research (WOTRO) for financially supporting my research work, to Jan Vermaat, my supervisor, for his assistance at the beginning and continuous guidance throughout the PhD endeavor. Jan Vermaat has provided his supervision with full responsibility not only when he was at the UNESCO-IHE but also after he left for the IVM in Amsterdam. I have learnt a lot from his precious ideas and suggestions, particularly on data analysis and demonstration of research outputs. Without his helpfulness, I would not have passed through the difficult circumstances.

I am also grateful to Prof. Dr. Patrick Denny, my promoter, for accepting me as his student and for critically reading the manuscripts, to Erik de Ruyter van Steveninck for his general assistance during my visits to UNESCO-IHE, to Erick de Jong for processing all academic formalities professionally, to Dr. Somsak Boromthanarat, CORIN director, for giving me permission for this sandwich PhD as well as for providing vehicles and equipment for field work, to Carlos Duarte, CERDS coordinator, and Jorge Terrados, for fruitful discussions on experimental design and data analysis, and for cooperation in the field work and writing of papers.

Field experiments and *in situ* data collection in the difficult working environment of the mangrove forest would not have been possible without the magnificent assistance of Nawee Noon-anand, Chamnan Worachina, Phollawat Saijung, Warut Rattananond and the fishermen at Pak Phanang West village. For all: I greatly appreciate having worked with you.

My acknowledgement also goes to UNESCO-IHE staff: Frank Wiegman for laboratory work, Vandana Sharma, Jolanda Boots and Ineke Melis for VISA and accommodation arrangement. I am thankful to Dr. Sakda Jongkaewwattana at Chiang Mai University, Johan van de Koppel, NIOO, Yerseke and Florian Eppink, IVM-VU, for useful advice in modeling. I thank my colleagues Dr. Nipa Panapitukkul, Dr. Ayut Nissapa, and Dr. Naiyana Srichai for discussion and encouragement.

Finally, I would like to thank my parents, sister, brothers, nieces and nephews, for their love and encouragement which helped pushing me passed through this long journey.

Udomluck Thampanya
September 2005

Summary

Mangroves are a specific type of evergreen forest that is found along the coastlines of tropical and subtropical regions, particularly along deltas and bays where rivers discharge freshwater and sediment to the sea. These mangroves provide important ecological and socio-economic functions to coastal dwellers and societies. For example, they are natural spawning and living ground for many species of fish and crustaceans. Their timber benefits local people as construction material, firewood and charcoals and their marvelous root systems contribute to sediment deposition, mud flat formation and substrate stabilization. Thus, these forests also function as a shelter belt during storms, cyclones and tsunamis. This function was evident when the mangroves in Southern Thailand helped to mitigate the recent 2004 tsunami's devastation on lives and properties of the villages situated behind them.

The present status of these valuable resources is critical, particularly in Southeast Asia, where the mangrove forest area is estimated to have declined by more than 50 percent over the past 35 years. The major causes of this loss were encroachment due to population expansion, conversion to aquaculture ponds, coastal erosion, lack of awareness and obscure or poorly enforced regulations. In Thailand, for example, recent satellite images show ample evidence of established and widespread aquaculture ponds along the coast of the Gulf of Thailand due to lack of restriction. More recently, government and public awareness of the importance of mangroves has risen, which has translated into several restoration and afforestation projects. Restoration of degraded mangrove areas was found to require much effort and encountered land-right problems. Therefore, many projects had switched to afforest on newly formed mudflats. Such projects, however, have met with variable success or failed to achieve the stated goals. This was probably because planting was carried out straightaway without carefully determining site-species suitability, appropriate planting technique as well as self-recovery or self-colonization capabilities from nearby stands.

To obtain useful information for supporting sound mangrove and coastal zone management, this dissertation aims to gain a better understanding of mangrove (re)-colonization and factors effecting colonization success as well as factors underlying coastal changes. Firstly, the impact of sedimentation and water turbulence on seedling survival and growth of three common SE Asian mangrove taxa: *Avicennia*, *Rhizophora* and *Sonneratia*, were examined experimentally at the bay scale. Then, this research was broadened to assess the coastal dynamics of Southern Thailand by synthesizing data from coastal surveys over the period 1961-2000 along with field studies. Finally, simple demographic models were developed to simulate the mangrove colonization process.

The two experiments reveal that after successful establishment on the mudflat, seedlings were susceptible to mortality due to sudden high sediment burial and water turbulence. Mortality was found to increase with increasing sediment accretion and none of *Avicennia* seedlings was able to survive under the highest experimental burial of 32 cm while those of *Rhizophora* and *Sonneratia* still showed substantial survival (30% and 60%, respectively). Among the three taxa, Sonneratia was the least affected due to burial and exhibited a rapid growth rate. Seedlings of *Rhizophora* survived less in highly exposed plots (low neighboring plant density) than in sheltered ones. In contrast, survival of *Avicennia* and *Sonneratia* seedlings were higher at more exposed plots. This finding confirms the behavior of *Avicennia* and *Sonneratia* as pioneer species that may colonize unoccupied mudflat and *Rhizophora* as a successor. It also suggests a higher success of the first two species in re-colonization. However, in areas where sudden high sediment loads are possible, *Sonneratia* might be better able to cope with burial than *Avicennia*.

The assessment of changes along Southern Thai coastlines revealed that the coast of the Gulf of Thailand had undergone more change than the western side of the peninsula. Ongoing erosion occurred irregularly at the high energy coastal segments and was observed along 29% and 11% of the total coastal length of the east and the west coasts, respectively. Subsequent total area loss accounted for 116 ha annually (91 and 25 ha for the east and the west coasts, respectively). Factors responsible for coastal erosion were found to be mangrove area loss, increase of shrimp farming area, less sediment delivery to coastal area due to upstream river damming and exposure to fetch from the monsoonal wind. Erosion was less in mangrove-dominated coastlines, and progression occurred mainly in sheltered and mangrove-dominated coastal segments. Areal progress accounted for 37 ha per annum on the east and 5 ha on the west coasts. Thus, summed over the whole of Southern Thailand, a net erosion of 74 ha y-1 was observed over the past three decades. These results also reveal that the existence and progression of the mangroves as well as sufficient sediment supply to coastal area contributed significantly to coastal stability.

Thirty-year runs of the demographic simulation model revealed that herbivory and water turbulence were the main factors influencing the success of colonization in a mangrove-dominated bay while gradual sedimentation and salinity had little effect. *Avicennia* was the most successful taxon in colonizing open mudflats followed by *Sonneratia* while *Rhizophora* exhibited less success. However, colonization success may vary with changes in environmental conditions as revealed by our predictive scenarios. For example, accelerated sea level rise will reduce the success in colonization of all three taxa and lack of freshwater discharge due to river damming may enhance the colonization success of *Avicennia* but adversely affect *Sonneratia*.

Overall, this study suggests that success in re-colonization of mangroves depends on both ecological and hydrological factors. Seedling herbivory and water turbulence are important factors that may seriously hinder colonization success. Also, the three taxa studied were found to differ critically in colonization capacity. Gradual sedimentation has little effect on mangrove colonization but positive sedimentation provides habitats for mangrove to colonize, particularly, in sheltered coastal segments. Less sediment delivery due to upstream river damming associated with conversion of mangrove and beach forest to aquaculture ponds will probably intensify coastal erosion in vulnerable areas. Therefore, managers should pay more attention to the balance between ecological and socio-economic demands for sustainable development of the coastal zone.

Table of Contents

Chapter 1

Propagules, seeds and dispersal of mangroves and their potential sensitivity to sedimentation, a general introduction

Introduction

This dissertation attempts to analyze the effects of sedimentation on mangrove colonization. It does so in relation to other factors that may be of importance. Before the main body of this dissertation, this general introductory chapter deals with the overall biology of mangroves in as much as that is relevant for dispersal and colonization. The term "colonization success" here is used in the sense that a mangrove seedling has succeeded in establishment on the mudflat substrate and subsequently survived until the adult tree stage.

Mangroves and their importance

In the early 1900's, spatial extent of mangroves in the world was estimated to be approximately 170,000 km^2 with 35% found in SE Asian region (Aksornkoae 1993). There are more than 70 species of true mangroves worldwide of which approximately 65 species contribute significantly to mangrove forest structure (Field 1995). With 40 species, the Indo-Pacific region exhibits the greatest diversity, while 15 and 10 species were found in Africa and the Americas, respectively. In Thailand, the estimated mangrove area in 2000 was 2,062 km^2 (RFD 2004) or approximately 1.2% of the world mangroves with the dominant species belonging to the genera *Rhizophora*, *Ceriops* and *Bruguiera* in the family Rhizophoraceae, *Sonneratia* in the family Sonneratiaceae and *Avicennia* in the family Avicenniaceae (Aksornkoae 1993).

Presently, the importance of mangroves both in terms of ecological and economic values is widely recognized (Aksornkoae 1993; Plathong & Sitthirach 1998; Semesi 1998; Janssen & Padilla 1999). Mangroves' diverse and numerous root systems in combination with detritus and nutrient abundance provide excellent shelter, feeding and spawning grounds for many aquatic organisms (Rönnbäck 1999). In Thailand, 72 species of fish, 50 species of shrimp and 30 species of crab are found in or adjacent to the mangrove areas (Aksornkoae, 1993). Abundant commercial species include mullets (*Mugilidae*), milkfish (*Chanos chanos*), sea bass (*Lates calcarifer*), penaeid shrimps (*Penaeus monodon)* and mud crabs (*Scylla serrata*). Therefore, mangroves have been important resources and coastal populations depend strongly in their livelihood both on coastal fisheries and mangrove production. Reportedly, there are over 120 species of fish caught annually in the world's largest mangrove area of the Sundarbans (FAO 1994). Mangroves also provide important habitats for charismatic species including reptiles, monkeys and waterfowl and shorebirds such as kingfishers, bee-eaters, sea eagles and egrets. In addition, mangroves are considered to play an important role in coastline protection as a buffer against storm-tide surges (Mazda et al. 1997).

Mangrove adaptations

Growing at the transition between land and sea, mangroves have developed special anatomical characteristics to survive in such uncertain habitats, for examples, aerial roots. Subtle variations in key environmental factors have resulted in further adaptations not only among species, but also between individuals of the same species living in different conditions. Differences in climate, geomorphology, hydrodynamic disturbances and sedimentation regime have created differential incentives for root characteristics such as strength, retention of oxygen, nutrient acquisition, and resilience to sedimentation.

Root systems of some species are probably better as anchorage while those of others are better as means to acquire nutrients from the sediment, or oxygen from the air. For instance, stilt or prop roots of *Rhizophora* offer support to tall trees to withstand the forces of strong winds (Figure 1.1a; Field 1995). For *Avicennia* and *Sonneratia* which grow at lower tide levels, pneumatophores aim to acquire atmospheric oxygen (Figure 1.1b, 1.1c; Field 1995). From the study of mangrove root systems, Komkris (1993) reported that periderms and cortex of stilt root of *Rhizophora* contain a lot of tannin that contribute to the solidity of roots whilst pneumatophores of *Avicennia* consist of aerenchymatous tissue containing many air spaces for oxygen reserve. However, lenticels are an important component for aeration in all types of aerial roots (Tomlinson 1986). A lenticel is an elongated mass of loose cells protruding through a fissure in the periderm (Hutchings & Saenger 1987). Lenticels are conspicuous on pneumatophores of *Avicennia* and *Sonneratia* as well as on the stilt root and hypocotyl of *Rhizophora* (Youssef & Saenger 1996; Ashford & Allaway 1995). Some mangrove species are able to respond to inundation or sedimentation. The stilt roots of *Rhizophora* are capable of elongating up to eight meters (Santisuk 1983). From the observation of Aksornkoae (1975), it was found that the height of aerial roots of *Rhizophora* in a wide tidal range area was higher than that in a short tidal range. Additionally, pneumatophores of *Avicennia* have limited height of less than 30 cm and develop little secondary thickening while those of *Sonneratia* are taller because they have a much longer period of development and the highest pneumatophores ever found was three meters (Tomlinson 1986). Therefore, *Avicennia* trees are not likely to survive under abrupt sediment accretion of more than 30 cm.

Figure 1.1. Mangrove root characteristics: (a) network of stilt roots growing from the trunk of *Rhizophora mucronata* and extending downward (b) numerous pnumatophores of *Avicennia alba* emerging out of the mud, and (c) *Sonneratia caseolaris* tree showing dense and extensive pnumatophores.

Seeds and seedlings

Tide and wave action along with substrate instability all challenge successful mangrove regeneration. Different species have adopted strategies to enhance successful recruitment that vary in terms of seed development, size, and number. Seeds of some species germinate while they are still on the tree for the best chance of regeneration; a strategy called vivipary. The embryo of this type of seeds has no dormancy but grows out of the seed coat and fruit while it is still attached to the parent tree. This can be found mostly in the family Rhizophoraceae (Figure 1.2a; Tomlinson 1986). After detaching, the agent of dispersal (propagule) of these species is a seedling. A variation is crypto-vivipary, in which the embryo emerges only from the seed coat but does not grow sufficiently to rupture the pericarp. At this level of development, the propagule is more accurately considered a fruit, rather than a seedling. Crypto-vivipary is found only in some taxa such as *Avicennia* (Figure 1.2b) and *Nypa* (Hutchings & Saenger 1987). Both vivipary and crypto-vivipary seedlings contain nutrients storage, which may assist rapid rooting in the unstable environment. Alternately, non-vivipary mangrove species rely on the traditional strategy of producing seeds, rather than precociously developed seedlings. Examples are *Xylocarpus*, in which the fruit splits on the parent tree to release its angular seeds, and *Sonneratia*, which produces a fruit containing numerous tiny seeds (Figure 1.2c; Tomlinson 1986).

Figure 1.2. Mangrove diaspores:

(b) cluster of
Avicennia alba
seedlings

(a) seedlings of *Rhizophora mucronata*   and (c) fruit of *Sonneratia caseolaris.*

In general, there appears to be an inverse relationship between size and number of propagules produced. Species with smaller propagules such as *Avicennia* have higher seed production than those with larger propagules. From the study of Thammathaworn (1982), it was found that one main branch of *Avicennia alba* contained more than 150 mature fruits. Additionally, from the observation of mangrove seed production at Pak Phanang Bay (Southern Thailand) during the year 1997-1999, the author found that one main branch of eight years old *Avicennia alba* tree contained approximately 300 fruits while *Rhizophora mucronata* of similar age produced approximately 10-25 seedlings per branch. It was also found that *Sonneratia caseolaris* at similar age produced approximately 10-70 fruits per branch with each fruit containing around 500-1600 tiny seeds.

The success of mangrove fruit development varies between species. Mangroves reportedly produce large amounts of seeds or seedlings (Tomlinson 1986). Considerable mortality often occurs in fruit development. Because of its economic value, a number of studies have considered the reproductive success of *Rhizophora* determining that the percentage of initial flower buds to reach the mature fruit stage is low. Gill & Tomlinson (1971) reported that less than 7% of *Rhizophora mangle* flowers produced mature propagules. Christensen & Wium-Andersen (1977) studied phenology of *Rhizophora apiculata* on Phuket Island (Thailand) and found that only 7% of flower buds formed flowers and only 1-3% of the buds formed fruits. The study of Kongsangchai & Havanond (1985) at Klang,

Malaysia reported that *Rhizophora apiculata* produced only one mature propagule from 194 flower buds and *Rhizophora mucronata* produced seven mature propagules from 428 flower buds, resulting in approximately 99% abortion rate for both species. Despite these high fruit mortality rates, habitat condition, probably tidal and wave buffeting, is thought to be the most important factor in determining seedling success (Clarke 1995).

Fruit falling times

Flowering and fruiting of mangroves can be observed throughout the year, particularly in Rhizophoraceae (Chapman 1976; Tomlinson 1986; Aksornkoae 1993). Continuous reproduction is probably an effort to maximize the amount of fruits available for mangrove regeneration. Fruit falling at an inappropriate time, however, may reduce this reproductive achievement. Furthermore, detachment from the parent tree and buoyant floating dispersal is limited to rather brief period of a few months.

In Thailand, the falling time of mature fruits or seedlings varies among species and geographic locations (Table 1.1). More importantly, phenological characteristics of mangroves, like others vegetation, depend largely on environmental factors such as rainfall and temperature. Observations of fishermen at Pak Phanang Bay, SE Thailand, for example, suggest that *Avicennia* fruit production was considerable in the wet year and these fruits fell from the mature tree in normal falling time (November-January, Table 1.1) whereas in the dry year, the fruit production was less and the falling time was somewhat later than usual. Information on fruit dislodgement is useful for mangrove plantation because it suggests when sources of seedlings are available and which times are most suitable for planting.

Table 1.1. Normal period of mature fruit/seedling falling of some mangrove species in the west and the east coasts[a] of Thailand.

species	location	Jan	Feb	Mar	Apr	May	Jun	Jul	Aug	Sep	Oct	Nov	Dec
Rhizophora apiculata	West				*	*	*	*					
	East	*									*	*	*
Rhizophora mucronata	West				*	*	*	*					
	East			*	*	*	*						
Avicennia alba	West								*	*			
	East	*										*	*
Avicennia officinalis	West									*	*		
	East	*										*	*
Sonneratia caseolaris	West								*	*			
	East		*	*									
Rainfall period	West					————————————							
	East	—									————————————		

[a]Observations from – the West coast: Ranong, Phuket, Phang Nga and Krabi; the East coast: Nakhon Si Thammarat, Songkhla and Pattani. Sources: Christensen & Wium-Andersen 1977; Thampanya (pers. Obs.).

Mangrove propagule and establishment

All mangrove propagules are buoyant at release allowing them to float on water (Tomlinson 1986; Clarke & Myerscough 1991; Clarke et al. 2001). When a mature fruit or seedling falls from the tree it might establish itself on adjacent mud if exposed, or be dispersed by currents (Tomlinson 1986). Variation in sizes and salt concentrations of propagules probably leads to differences in floating time and establishment among species. The smaller propagules such as in *Avicennia* (Table 1.2) are released in high numbers and they are easily dispersed by water while the larger propagules such as in *Rhizophora* are released in small numbers but can sustain a much longer period of time waiting for an appropriate condition to establish (Field, 1995). According to the study of Rabinowitz (1978), the floating time could range from a day for *Pelliciera* to an unlimited period for *Avicennia germinans* and longevity of propagules can vary from 35 days for *Laguncularia* to more than a year for *Rhizophora mangle*.

Table 1.2. Approximate quantification of reproduction traits in some common SE Asian mangrove species.

species	propagation unit[a]	length (cm)[a]	width or diameter (cm)[a]	no. of seeds/fruit[b]	no. of fruits/branch[c]
R. apiculata	seedling	25 - 50	1.5 - 3.0	1	-
R. mucronata	seedling	30 - 70	2.0 - 4.0	1	10-25
A. alba	fruit	2.0 - 4.0	1.2 - 1.8	1	300 (approx.)
A. officinalis	fruit	2.0 - 2.5	3.0	1	-
A. marina	fruit	2.0 - 3.0	2.0 - 3.0	1	-
S. alba	fruit	3.0 - 3.5	3.0 - 4.7	500 (approx.)	-
S. caseolaris	fruit	3.0 - 4.0	3.0 - 5.0	500 – 1600	10-70

Sources: [a]Tomlinson (1986), [b]Komkris (1993), [c]Thampanya (pers. Obs.).

Mangrove propagules usually anchor horizontally after loosing buoyancy. The common view is that mangrove trees are similar to other water-dispersed trees which have a "stranding strategy": to reach a vertical position rapidly so that interference with gas exchange by water is minimized (Tomlinson 1986; Tomlinson & Cox 2000). For example, after the *Rhizophora* propagule touches the ground, initial roots will be developed especially on the lower surface of the hypocotyl within approximately 3-5 weeks, the distal seedling will become erect (Figure 1.3a) by an internal mechanism with the development of tension wood fibres in secondary xylem (Tomlinson & Cox 2000).

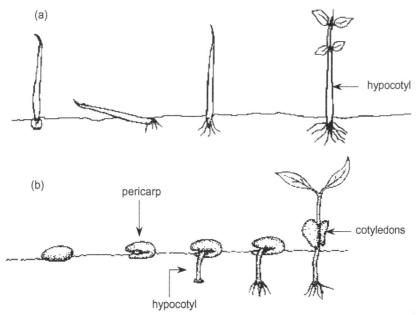

Figure 1.3. Stranding of: (a) *Rhizophora* seedling and (b) *Avicennia* seedling (after Komkris, 1993).

For smaller propagules, such as that of *Avicennia*, the establishment process is also complicated (Figure 1.3b). The propagule will sink approximately four days after splitting of the pericarp and will root as soon as it becomes stationary (Tomlinson 1986). First, the hypocotyl elongates out from the seed followed by root development from the tip of hypocotyl. Finally, the cotyledons split and the plant becomes erect. Approximate times for the seedling to become firmly rooted are five days for *Avicennia marina* and seven days for *Avicennia germinans* (Hutchings & Saenger 1987).

Successful recruitment is highly related to site conditions, but may also be influenced by species-specific adaptations. The role of adaptation can be seen by comparing the establishment of *Avicennia* and *Rhizophora*. Once the propagule touches the ground, the time required for establishment of the smaller propagules is shorter than that of the larger. This may provide an advantage to smaller propagules to establish in more turbulent areas.

Mortality, however, often occurs in both small and large propagules either during dispersal or after establishment. Propagules can be attacked by insects or larger predators at any time. After release, some may be lost in the water and get flushed out to the sea whereas others might succeed in attaching to the ground. However, during subsequent establishment, mortality may be substantial, due to, e.g., strong water movement, drought or salinity stress and herbivory. In a study of establishment and mortality of propagules in permanent plots in Australia, Hutchings & Saenger (1987) found that mortality rates of *Rhizophora, Ceriops* and *Avicennia* were variable and site-dependent (Figure 1.4).

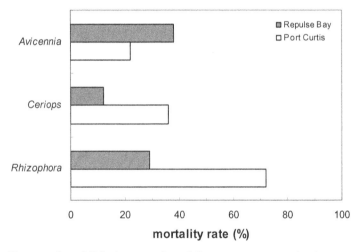

Figure 1.4. Mortality rate of established propagules of three mangrove species in permanent plots at Port Curtis and Repulse Bay, in Australia (cf. Hutchings & Saenger 1987).

Mangrove propagules are well known to be capable of overseas dispersal (Tomlinson 1986), but the quantification of propagule fluxes and establishment success has been attempted only rarely (Hutchings & Saenger 1987). Also, mangrove forestry and demographic modelling has largely focused on established saplings (Berger & Hildenbrandt 2000; Berger et al. 2002). Since re-afforestation programs often have failed in the past, it was felt that comparative, *in situ* study of mangrove seedling establishment would be opportune and useful.

Sedimentation and mangroves

Coastal sedimentary processes are inherently dynamic but major changes have been related to anthropogenic activities in upland areas (Syvitski et al. 2005). Considerable amounts of terrestrial sediments are transported by rivers to coastal waters annually. For instance, approximately 47 million cubic meters per annum of suspended solids is delivered into the Gulf of Thailand (Sriratanatabucanon 1991). In southern Thailand, approximately one million ton of sediments is discharged to Pak Phanang Bay by Pak Phanang river annually (JICA 1987). Increased sedimentation by an order of magnitude has been reported for SE Asia during past decades (Milliman & Syvitski 1992). Vast deforestation of upland and conversion of mangrove forests due to rapid population expansion and economic development are seen as important causes of increased sediment loads of this region (Aksornkoae 1993).

Sedimentation is considered an important factor in mangrove ecology (Hutchings & Saenger 1987; Robertson & Alongi 1992; Clarke 1995; Ellison 1998) as mangroves usually grow at low-lying continental coastlines where riverine sediments are delivered to the sea. Thus, sedimentation is inevitably paramount importance to these coastal forests. Its impacts can be positive or negative. On the positive side, sediment accretion may open new habitat for mangroves to colonize (e.g. Purba 1991; Lee et al. 1996; Panapitukkul et al. 1998). In contrast, sudden high sedimentation events may occur during heavy rainfall and can cause reduction in growth and mortality of mangroves if their aerial roots are blocked (Hutchings & Saenger 1987; Terrados et al. 1997; Ellison 1998). Under normal conditions, however, sediments may deposit at a rate of up to 1.5 cm per year allowing the mangrove root system to respond and acclimate, so that aeration can continue (Hutchings & Saenger 1987). Additionally, reduction in sediment delivery to coastal waters as a result of river damming also have negative impacts to coastal environment. Sediment starvation of the littoral zone due to reduced sediment delivery may lead to erosion in susceptible areas including mangrove-dominated coasts (ISME/GLOMIS 2002; Batalla 2003; Bird et al. 2004).

In Thailand, the coastlines have experienced changes both in terms of accretion and erosion. Several coastal segments have undergone extensive progression due to sediment accretion whilst erosion also occurred elsewhere (Vongvisessomjai et al. 1996), particularly, along the Gulf coastline where used to be mangrove-dominated. In addition, growing awareness of mangrove degradation and the importance of their ecosystems has created demand for mangrove rehabilitation and afforestation programs in the country (Kongsangchai 1995; Havanond 1995). Apart from degraded mangrove areas, open mudflats were usually selected as plantation sites. As mentioned earlier, these accreting mudflats have the potential to be colonized by pioneer mangrove species from nearby stands, therefore, promoting or enhancing colonization in such areas is probably another alternative way of economically mangrove rehabilitation. More importantly, the existence of mangroves along the coast may help reducing coastal erosion (Mazda et al. 1997). This, however, further requires sound understanding of the dynamic ecological processes that underlay successful mangrove colonization, the impacts of sedimentation on mangroves and the behaviour of mangrove-existing coastlines.

Structure of the dissertation

Regarding the above requirement, this thesis focuses on the colonization success of mangroves and the impacts of sedimentation on these important coastal resources. The dissertation quantitatively assesses the impacts of sedimentation and exposure on establishment success of seedlings of three common Thai mangrove taxa, *Avicennia*, *Rhizophora* and *Sonneratia*. It verifies and, thereafter, uses spatial age distribution patterns to estimate coastal accretion rates and offsets these to overall coastal accretion and erosion in the region. Finally, these results and data obtained from literature are combined into simple demographic models of seedling establishment success that have been calibrated and verified.

The dissertation is composed of seven subsequent chapters. Chapter 1 (this chapter) has gathered basic information necessary for an understanding of mangrove ecology. In Chapter 2, an *in situ* experiment was carried out to elucidate which species successfully colonized on the accreting mudflats of Pak Phanang Bay, southern Thailand, and the effect of variable exposure is described. Thereafter, the likelihood of success of the three species in habitats under different sediment accretion levels was examined by *in situ* experiment sediment burial (Chapter 3). Diameter and height-age regression equations for common Thai mangrove species are developed in Chapter 4. These results are used as tools for estimating age of mangrove stands in the following Chapter 5, which assesses dynamics of Southern Thai coastlines by synthesizing data of coastal surveys together with field studies and remotely sensed data analysis. This chapter also examined the behavior of mangrove-dominated coastline in response to sediment accretion and reduction. In Chapter 6, colonization success of the three mangrove species has been integrated in demographic models of the recruitment from seedling to sapling and tree stages. The concluding chapter, a general discussion focuses on useful recommendations for coastal management and mangrove rehabilitation (Chapter 7).

References

Ashford, A.E. and Allaway, W.G. 1995. There is a continuum of gas space in young plants of *Avicennia marina*. Hydrobiologia 295: 5-11.

Aksornkoae, S. 1975. Structure regeneration and productivity of mangrove in Thailand. Ph.D. Thesis, Michigan State University, USA.

Aksornkoae, S. 1993. Ecology and Management of mangrove. IUCN, Bangkok, 176 pp.

Batalla, R.J. 2003. Sediment deficit in rivers caused by dams and instream gravel mining. A review with examples from NE Spain. Cuaternario y Geomorfología 17: 79-91.

Berger, U. and Hildenbrandt, H. 2000. A new approach to spatially explicit modelling of forest dynamics: spacing, aging and neighbourhood competition of mangrove trees. Ecological Modelling 132: 287-302.

Berger, U., Hildenbrandt, H. and Grimm, V. 2002. Towards a standard for the individual-based modeling of plant populations: self-thinning and the field-of-neighborhood approach. Natural Resource Modeling 15: 39-54.

Bird, M., Chua, S., Fifield, L.K., Teh, T.S. and Lai, J. 2004. Evolution of the Sungei Buloh-Kranji mangrove coast, Singapore. Applied Geography, 24: 181-198.

Chapman, V.J. 1976. Mangrove Vegetation. Vaduz: Cramer, 480 pp.

Christensen, B. and Wium-Andersen, S. 1977. Seasonal growth of mangrove trees in Southern Thailand I. The phenology of *Rhizophora apiculata* Bl. Aquatic Botany 3: 281-286.

Clarke, P.J. and Myerscough, P.J. 1991. Buoyancy of *Avicennia marina* propagules in south-eastern Australia. Australia Journal of Botany 39: 77-83.

Clarke, P.J. 1995. The population dynamics of the mangrove *Avicennia marina*; demographic synthesis and predictive modelling. Hydrobiologia 295: 83-88.

Clarke, P.J., Kerrigan, R.A. and Westphal, C. 2001. Dispersal potential and early growth in 14 tropical mangroves: do early life history traits correlate with patterns of adult distribution. Journal of Ecology 89: 648-659.

Ellison, J.C. 1998. Impacts of sediment burial on mangroves. Marine Pollution Bulletin 37: 420-426.

Elster, C. 2000. Reason for reforestation success and failure with three mangrove species in Colombia. Forest Ecology and Management 131: 201-214.

Food and Agriculture Organization (FAO). 1994. Integrated management of coastal zones. FAO fisheries technical paper no. 327. FAO, Rome, 167 pp.

Field, C.D. 1995. Journey amongst mangroves. The International Society for Mangrove Ecosystem (ISME), Okinawa, Japan, 140 pp.

Gill, A.M. and Tomlinson, P.B. 1971. Studies on the growth of red mangrove (*Rhizophora mangle* L.). III. Phenology of the shoot. Biotropica 1: 1-9.

Gomez, E.D. 1988. Overview of environmental problems in the East Asian seas region. AMBIO 17: 166-169.

Havanond, S. 1995. Re-afforestation of mangrove forests in Thailand. In: Khemnark C. (eds), Ecology and management of mangrove restoration and regeneration in East and Southeast Asia. Proceeding of the ECOTONE IV. Kasetsart University, Bangkok, Thailand. pp. 203-216.

Hutchings, P. and Saenger. P. 1987. Ecology of mangroves. University of Queensland Press, 388 pp.

ISME/GLOMIS. River damming and changes in mangrove distribution. ISME/GLOMIS Electronic Journal, Volume 2, July 2002.

Japan International Corporation Agency (JICA). 1987. Report on the basic design for constructing the Nakhon Si Thammarat fishing port in the Kingdom of Thailand. Japan International Corporation Agency, 259 pp.

Janssen, R. and Padilla, J.E. 1999. Preservation or conservation? Valuation and evaluation of a mangrove forest in the Philippines. Environmental and Resource Economics 14: 297-331.

Komkris, T. 1993. The structure of mangrove. Kasertsat University Press. 151 pp. (in Thai).

Kongsangchai, J. and Havanond, S. 1985. Some phenological characteristics of *Rhizophora apiculata* and *Rhizophora mucronata*. Paper presented at the Seminar on Mangrove Studies, NODAI Research Institute, Tokyo University of Agriculture, Japan, August 1985.

Kongsangchai, J. 1995. Problems of mangrove degradation. In: Khemnark C. (ed), Ecology and management of mangrove restoration and regeneration in East and Southeast Asia. Proceeding of the ECOTONE IV. Kasetsart University, Bangkok, Thailand, pp. 110-128.

Lee, S.K., Tan, W.H. and Havanond, S. 1996. Regeneration and colonisation of mangrove on clay-filled reclaimed land in Singapore. Hydrobiologia 319: 23-35.

Mazda, Y., Magi, M., Kogo, M. and Hong, P.N. 1997. Mangroves as a coastal protection from waves in the Tong King delta, Vietnam. Mangroves and Salt Marshes 1:127-135.

Milliman, J.D. and Syvitski, J.P.M. 1992. Geomorphic/tectonic control of sediment discharge to the ocean: the importance of small mountainous rivers. Journal of Geology 100: 525-544.

Panapitukkul, N., Duarte, C.M., Thampanya, U., Terrados, J., Keowvongsri, P., Geertz-Hansen, O., Srichai, N. and Boromthanarat, S. 1998. Mangrove colonization: Mangrove progression over the growing Pak Phanang (SE Thailand) mud flat. Estuarine, Coastal and Shelf Science 47: 51-61.

Plathong, J. and Sitthirach, N. 1998. Traditional and current use of mangrove forest in Southern Thailand. Wetlands International-Thailand Programme/PSU, Publication No.3, 91 pp.

Purba, M. 1991. Impact of high sedimentation rate on the coastal resources of Segara Anakan, Indonesia,. In: L.M. Chou, T.E. Chua, H.W. Khoo, P.E. Lim, J.N. Paw, G.T. Silvertre, M.J. Valencia, A.T. White and P.K. Wong (eds), Towards an integrated management of tropical coastal resources. ICLARM conference proceedings 22: 143-152.

Rabinowitz, D. 1978. Dispersal properties of mangrove propagules. Biotropica 10: 47-57.

Robertson, A.I. and D.M. Alongi (eds.) 1992. Tropical mangrove ecosystems. Coastal and estuarine series 41. American Geophysical Union, Washington, D.C, 330 pp.

Royal Forest Department (RFD). 2004. Forest statistics 2002. Royal Forest Department 61 Phaholyathin, Ladyao, Bangkok, Thailand. Website: www.forest.go.th.

Rönnbäck, P. 1999. The ecological basis for economic value of seafood production supported by mangrove ecosystem. Ecological Economics 29: 235-252.

Santisuk, T. 1983. Taxonomy and distribution of the terrestrial trees and shrubs in the mangrove formations in Thailand. Nat. His. Bull. Siam Soc. 31: 63-91.

Semesi, A.K. 1998. Mangrove management and utilization in Eastern Africa. Ambio 27: 620-626.

Sriratanatabucanon, M. 1991.   State of Coastal Resources Management Strategy in Thailand. Marine Pollution Bulletin 23: 579-586.

Syvitski, J.P.M, Vörösmarty, C.J., Kettner, A.J and Green, P. 2005. Impact of humans on the flux of terrestrial sediment to the global coastal ocean. Science 308: 376-380.

Terrados, J., Thampanya, U., Srichai, N., Keowvongsri, P., Geertz-Hansen, O., Boromthanarat, S., Panapitukkul, N. and Duarte, C.M. 1997. The effect of increased sediment accretion on the survival and growth of *Rhizophora apiculata* seedlings. Estuarine, Coastal and Shelf Science 45: 697-701.

Thammathaworn, S. 1982. Botanical study on *Avicennia alba* Bl. In: Proceeding of the 4th National Seminar on Mangrove Ecology, pp. 275-288. National Research Council of Thailand. (in Thai).

Tomlinson, P.B. 1986. The botany of mangroves. Cambridge University Press, 418 pp.

Tomlinson, P.B. and Cox, P.A. 2000. Systematic and functional anatomy of seedlings in mangrove Rhizophoraceae: Vivipary explained? Bot J Linn Soc 134:215-231

Vongvisessomjai, S. Polsi, R., Manotham, C. and Srisaengthong, D. 1996. Coastal erosion in the Gulf of Thailand. In: Milliman, J.D., and Haq, B.U. (eds.), Sea level rise and coastal subsidence, Kluwer Academic Publishers, 131-150.

Youssef, T. and Saenger, P. 1996. Anatomical adaptive strategies to flooding and rhizosphere oxidation in mangrove seedlings. Australian Journal of Botany 44: 297-313.

Chapter 2

Colonization success of common Thai mangrove species as a function of shelter from water movement

Udomluck Thampanya, Jan E. Vermaat and Carlos M. Duarte

Abstract

Seedling survival and growth of the three common SE Asian mangrove species *Avicennia alba*, *Rhizophora mucronata* and *Sonneratia caseolaris* were quantified experimentally along two spatial gradients of shelter: (1) between two stations, at the inner and outer end of the sheltered Pak Phanang Bay (SW Thailand), and (2) for each station, among plots across a gradient of vegetation density from the mangrove forest edge inwards. Exposure to water movement, quantified as gypsum clod card weight loss, was found to vary more than fivefold among seasons, which contributed most of the explained variance (73%). Variation among plots was higher than that among the two stations: clod card loss ranged between 3.0 and 4.6 g d^{-1} among plots, whereas the grand means of the two stations were 3.4 and 3.7 g d^{-1}, respectively. These differences among stations and plots were comparable to the patterns found for mangrove seedling survival. Survival was high (80-93%) in most treatments in *Rhizophora*, with the exception of the most exposed plot (30%). In the other two species, overall survival was significantly less, but highest in the outermost plots with the lowest tree density. This pattern confirms the successional status of these three mangrove species. Seedling growth, expressed as height increase, was significantly reduced with increasing neighbouring tree density for *Avicennia* and *Sonneratia*, whereas *Rhizophora* showed an opposite pattern. Internode production of all three species was highest in the most exposed plots. Overall, relative growth rate, expressed as height increase, declined with the age of the seedlings.

Based on Thampanya, U., Vermaat, J.E. and Duarte, C.M. 2002. Colonization success of common Thai mangrove species as a function of shelter from water movement. Marine Ecology Progress Series 237: 111-120.

Introduction

Seedling establishment is a critical stage in the life cycle of most angiosperms (Silvertown 1982). The advanced developmental stage of viviparous mangroves seedlings, which develop while still attached to the mother tree in the Rhizophoreae genera is generally interpreted as adaptive (Tomlinson 1986) and facilitating rapid establishment through rooting (Hutchings & Saenger 1987). After a period of positive buoyancy, dispersed propagules sink or become stranded, and successful rooting is the first step towards seedling establishment (Hutchings & Saenger 1987; Tomlinson 2000). However, establishment may be hindered by tidal current and wave buffeting (Clarke 1995) or lack of shelter from water movement (Clough 1982). Consequently, large proportions (as high as 90%) of already rooted seedlings may fail to establish fully (Delgado et al. 1999). Furthermore, post-establishment mortality during the first year is also considerable: Hutchings & Saenger (1987) reported 22, 36 and 72% mortality in the first year for *Avicennia, Ceriops* and *Rhizophora stylosa* respectively. In this stage, possible mortality sources include herbivory, drought or salinity stress, insufficient light, damage from drifting objects, and strong water movement (Hutchings & Saenger 1987; Clarke 1995). Published relevant quantitative data on mangrove seedling survival and sources of mortality, however, are scarce (Hutchings & Saenger 1987; Clarke & Allaway 1993).

Our aim was to assess the effect of seasonal variation in exposure to water movement on the survival and growth of seedlings of three common SE Asian mangrove species at two spatial scales: (1) at a larger scale along a gradient from the river mouth to the sea in an enclosed bay (Pak Phanang Bay, Southern Thailand); and (2) at a smaller scale along a gradient of neighbouring plant density from the edge of an expanding mangrove forest perpendicular into the forest. In addition to their interest to provide information on factors affecting the early success of mangrove seedlings the goals set here are important to design successful large-scale mangrove rehabilitation schemes, such as those carried out at present in Thailand and neighbouring countries (Havanond 1995; Field 1995).

Materials and methods

Study area

Pak Phanang Bay (8° 25.11′ N 100° 09.18′ E) is a large, shallow bay, sheltered from the Gulf of Thailand by a long northwest pointing sandbar (Figure 2.1). The eastern side of the bay is largely occupied by mangrove forest with an area of approximately 90 km^2. The Pak Phanang River, 110 km long, discharges into the bay from the south. Tides are mixed-diurnal and ranged between 0.1-0.7 m in August 1998 to 0.7-1.3 m above the lowest low tide level in January 1999. Published average velocities are 0.6 m sec^{-1} for ebb currents and 0.8 m sec^{-1} for flood currents (JICA 1987). Annually, the area experiences three distinct seasons; hot-dry season (Feb-May), rainy season (Jun-Sep) and the highest rainfall period of monsoon season (Oct-Jan, Figure 2.2), with an average annual rainfall of 2,400 mm. The mean air temperature is 28°C and the water temperature rages between 25°C-36°C. The experiment was conducted at the bay side edge of the mangrove forest. Two stations with different exposure were selected: station 1, at the mouth of Pak Kwang canal near the river mouth, and station 2 at the mouth of Ai Ho canal 5 km away from station 1. At each station, three plots of 10 m x 30 m were set up at different densities of naturally occurring, neighbouring plants: low, medium and high (Table 2.1). The three plots of station 1 were mostly occupied by *Sonneratia caseolaris* trees (2-5 m in height), saplings and seedlings with a few *Avicennia* (*Avicennia alba* and *Avicennia officinalis*) saplings and seedlings while all the plots of station 2 were occupied by *Avicennia* trees, saplings and seedlings.

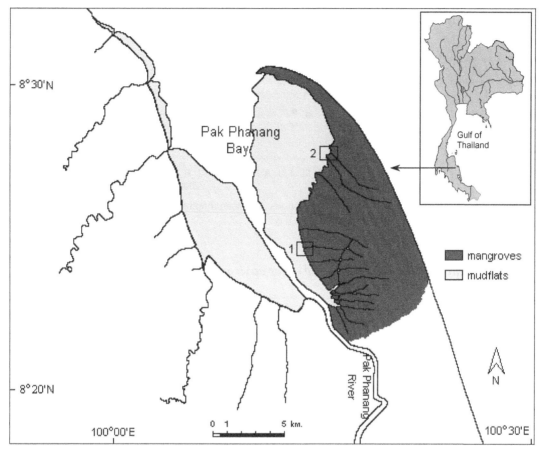

Figure 2.1. Map of Pak Phanang Bay, southern Thailand showing the two stations: Station 1 = Pak Kwang Station 2 = Ai Ho.

The upper areas and hinterland are occupied by planted *Rhizophora apiculata* and *Rhizophora mucronata* (5-20 m in height). Soil textures of both stations are silty-clay (53% clay for station 1 and 56% clay for station 2) and the substrate surface is relatively flat (+1 m elevation above MSL for both stations).

Table 2.1. Densities (individual per 100 m^2) of neighbouring plants existing in the low, medium and high density plots in two stations in Pak Phanang Bay. Mangrove plants were classified as seedlings (<1.5 m), saplings (between 1.5 and 2.5 m) or trees (>2.5 m).

station	plot	seedling	sapling	tree
Pak Kwang	low	97	5	2
	medium	253	24	13
	high	389	76	18
Ai Ho	low	68	3	1
	medium	282	26	11
	high	338	41	24

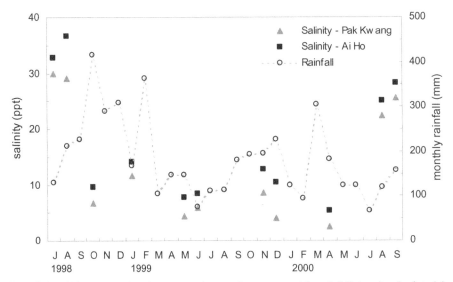

Figure 2.2. Salinity (‰) measured at the two stations and mean monthly rainfall (mm) calculated from the meteorological stations adjacent to Pak Phanang Bay.

Measurement of water movement and other environmental parameters

The dimensionless water movement was quantified using the dissolution rate of clod cards (Doty 1971; Jokiel & Morrissey 1993). This inexpensive and practical method quantifies weight loss of a sparingly soluble substance such as plaster of Paris or gypsum. Reportedly, weight loss is directly related to water motion (Petticrew & Kalff 1991; Jokiel & Morrissey 1993; but see Porter et al. 2000 for a critical assessment). A large number of plaster of Paris clods of considerable size (30-40 g) were produced using ice-cube trays as templates and allowed to air-dry at room temperature (28-33°C) for one week to reach constant weight. Thereafter, each clod was glued to a card of 5 cm x 7 cm rubber floor cover. After being oven-dried at 60°C for 24 h, the clod cards were pre-weighed before deployment. At the study site, each pair of clod cards was attached, using rubber elastic rings, to a piece (8 cm x 16 cm) of thin board with a hole punched in the middle (Figure 2.3). Fifteen pairs of clod cards were deployed randomly in each plot at both stations. The clod card set was fixed to the substrate with a stick. After being deployed for 24 h, the clod cards were collected and carefully carried back to the lab. Subsequently, they were allowed to air-dry for a week and were re-weighed after drying at 60°C for 24 h. Clod card deployment occurred randomly in the daily and monthly (lunar) tidal cycle: low tide occurred at 15 h, 18 h, 8 h, 8 h, 23 h, 5 h and 22 h, respectively, on subsequent days of deployment.

Figure 2.3. A pair of clod cards (a) and clod card deployment in a study plot (b).

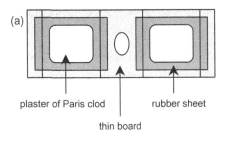

Data were analyzed using an ANOVA–GLM (General Linear Model) full factorial model comparing weight loss per day among seasons, stations and plots. Besides, closed boxes (15 litres) containing control clod cards were also deployed *in situ*, allowing the calculation of the diffusion factor (DF), *sensu* Jokiel & Morrissey (1993), as weight loss *in situ* divided by weight loss in still water. Salinity was measured *in situ* with a WTW conductivity meter. Rainfall data were obtained from the meteorological stations surrounding the study area.

Seedling survival and growth

Seedlings of *Avicennia alba, Sonneratia caseolaris* and *Rhizophora mucronata* were transplanted in May 1999, at the end of the hot-dry period. In each plot of both stations, thirty seedlings of each species were transplanted with 2 m x 2 m spacing. These transplanted seedlings were in the post-cotyledonary phase during which their survival appears to be largely resource dependent, *i.e.* independent of nutritional support from the parent (Clarke 1995). The seedlings of *Avicennia alba* and *Sonneratia caseolaris* were obtained by carefully shovelling up newly established natural seedlings in an adjacent area while the seedlings of *Rhizophora mucronata* were collected under mature trees upstream. The transplanted seedlings were tagged after replanting and recovered after a month. We then measured the height and number of internodes on the main stem, and repeated these measurements at approximately three-month intervals until the plants were more than one year old (Sep 2000). However, due to the very high monsoon water levels in Dec 1999 to Feb 2000 the monitoring of still submerged *Rhizophora* and *Avicennia* seedlings could not be realized. Only *Sonneratia* seedlings, which had grown fast enough to reach the water surface, were monitored during that time. Since there were slight differences in initial seedling sizes (with a mean height of 36, 37 and 16 cm for *Avicennia*, *Rhizophora* and *Sonneratia*, respectively), data on seedling height were transformed into relative growth rates (Hunt 1982) prior to analyses. At each visit we also recorded the number of surviving seedlings and analyzed seedling survival using a factorial GLM. Since growth was measured on the same experimental units (*i.e.*, the seedlings) a repeated-measures design was used to properly separate effects of time from treatment in ANOVA analysis (Potvin et al. 1990).

Results

Water movement

There was a significant difference in clod card dissolution between the two stations ($p < 0.001$, Table 2.2). Average clod card weight loss per day at Ai Ho Canal was somewhat higher than at Pak Kwang Canal (3.78 ± 0.08 g d^{-1} vs. 3.46 ± 0.07 g d^{-1}). Clod card dissolution rates were also significantly different among plots of different neighbouring plant densities and among seasons ($p < 0.001$, Table 2.2). The plots with the lowest density had the highest clod card dissolution rates (4.38 ± 0.10 g d^{-1}) while those with the highest density had the lowest (3.16 ± 0.09 g d^{-1}). In general, most of the variation (73%) was accounted for by the seasonality term (Table 2.2), with the lowest clod card weight loss in July 1998 and the highest in Oct 1998 (Figure 2.4), a pattern coupled with the monsoon period (cf. Figure 2.2). Also the lower maximum clod card dissolution rates in Nov 1999, compared to those of Oct 1998 (cf. Figure 2.4), coincided with lower rainfall in that period. The larger-scale exposure effect (stations) explained only 1% of total variation while the smaller-scale exposure effect (neighbouring plant density) contributed 8%.

The diffusion factor (DF) was subsequently calculated but only from Jan 1999 onward, due to loss of control clod card boxes for the first three measurements. Average DF was highest at the most exposed plot of Ai Ho canal in Apr 2000 (9.24 ± 0.10) and lowest at the densest plot of Ai Ho canal in May 1999 (2.20 ± 0.03).

Table 2.2. Three-way ANOVA examining the effects of stations, neighbouring plant densities (plots) and seasons (months) on clod card weight loss (g d^{-1}). Presented are the degree of freedom (*df*), *SS*, the percentage of variance explained (factor *SS*/total *SS* x 100) and the level of significance (*p*).

factors	*df*	*SS*	% variance	*p*
Station	1	31	1	0.000
Plot	2	324	8	0.000
Month	6	2905	73	0.000
Station x Plot	2	34	1	0.000
Station x Month	6	39	1	0.000
Plot x Month	12	104	2	0.000
Station x Plot x Month	12	35	1	0.000
Residual	1147	513	13	
Total	1188	4002	100	

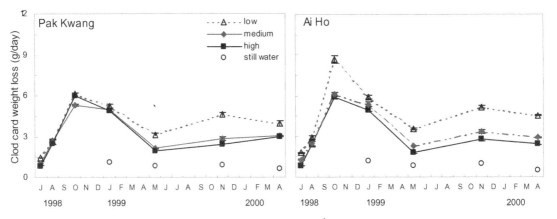

Figure 2.4. Seasonal pattern of clod card dissolution (g d^{-1}) at three plots of different neighbouring tree density (low, medium and high) and in still water of the two stations. Presented are means $\pm$ SE.

Seedling survival

Significant differences in seedling survival were detected among species and at the smaller-scale exposure (plot) level but not at the larger- scale exposure (station) level (Table 2.3). Most of the variation (47%) was explained by the difference in species while the difference in neighbouring plant density and the interaction between species and neighbouring plant density explained 11 and 13% of the total variation, respectively.

Table 2.3. Three-way ANOVA examining the effects of species, stations and neighbouring plant densities (plots) on seedling survival. Presented are the degree of freedom (df), SS, the percentage of variance explained (factor SS/total SS x 100) and the level of significance (p).

factors	df	SS	% variance	p
Station	1	32	1	0.394
Plot	2	518	11	0.006
Species	2	2186	47	0.000
Station x Plot	2	6	<1	0.948
Station x Species	2	7	<1	0.919
Plot x Species	4	599	13	0.019
Station x Plot x Species	4	35	1	0.934
Residual	30	1291	28	
Total	47	4613	100	

Survival curves of *Avicennia* and *Sonneratia* were comparatively similar but very different from those of *Rhizophora* (Figure 2.5). The numbers of surviving seedlings of the first two species declined throughout the experimental period, whereas those of *Rhizophora* remained fairly constant until after the monsoon season, when mortality was particularly apparent in the low density plot at Ai Ho. In *Avicennia,* mortality was spread evenly over the whole period, whilst *Sonneratia* experienced two periods of high mortality (Jun-Sep 1999 and Nov 1999-Feb 2000, Figure 2.5). Also, *Avicennia* and *Sonneratia* seedlings had higher survival rates in the plots with low plant density, *i.e.* at high exposure. In contrast, *Rhizophora* seedlings survived much better in the less exposed, denser plots (Figure 2.5). After a one year cycle, between 8 and 40% of the *Sonneratia* had survived, with survival being highest in the outermost low-density plots (Table 2.4). Likewise, *Avicennia* had the highest survival at the lowest density of neighbouring trees with the mortality rate ranging between 30-85%. In *Rhizophora,* mortality rates after one year were relatively low in the medium and high plant density plots (13 and 15%), but comparatively high in the low plant density plots (43%, Table 2.4).

Table 2.4. Mortality rate after one year of the three mangrove species in the plots of increasing neighbouring plant density. Presented are means ± SE calculated from both stations.

species	neighbouring plant density		
	low	medium	high
Avicennia alba	30.0 ± 0.0	71.7 ± 5.0	85.0 ± 8.3
Rhizophora mucronata	43.3 ± 26.7	15.0 ± 5.0	13.3 ± 6.7
Sonneratia caseolaris	60.0 ± 6.7	95.0 ± 5.0	91.7 ± 5.0

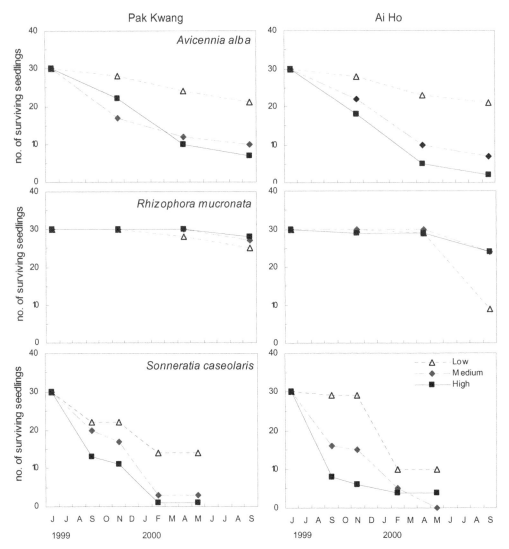

Figure 2.5. Survival curves of *Avicennia*, *Rhizophora* and *Sonneratia* seedlings at three plots of different neighbouring tree density at Pak Kwang and Ai Ho canals.

Seedling height

Firstly, the data of all three species were combined together and analyzed using repeated measures GLM to examine the effect on seedling growth of factors time, species, stations and plots of different neighbouring plant density. We found that the repeated measures factor (time) and its interaction with stations as well as its interaction with species and plots were all significant ($p<0.05$). We also found highly significant differences in relative growth rate (RGR_H) among species and between the two stations ($p<0.001$, Table 2.5) and slight significant differences among plots of contrasting density ($p<0.05$, Table 2.5). The *Sonneratia* seedlings had the highest RGR_H while those of *Avicennia* had the lowest (Table 2.6). Overall, RGR_H declined with increasing age of the seedlings (Figure 2.6).

Table 2.5. Result of repeated measures[a] GLM examining the effects species, stations and neighbouring plant densities (plots) on seedling height increase (RGR_H) for all three species and each species separately. Presented are the degree of freedom (*df*), sums of squares (*SS*) and the level of significance (*p*).

all species			
factors	*df*	*SS*	*p*
Species	2	54.907	0.000
Station	1	16.129	0.000
Plot	2	2.985	0.001
Station x Plot	2	6.661	0.000
Station x Species	2	17.866	0.000
Plot x Species	4	8.278	0.000
Station x Plot x Species	4	8.273	0.000
Within + Residual	222	42.245	

Avicennia			
factors	*df*	*SS*	*p*
Station	1	1.702	0.000
Plot	2	0.020	0.828
Station x Plot	2	0.774	0.002
Within + Residual	64	3.459	

Rhizophora			
factors	*df*	*SS*	*p*
Station	1	0.162	0.027
Plot	2	0.853	0.000
Station x Plot	2	0.081	0.288
Within + Residual	131	4.220	

Sonneratia			
factors	*df*	*SS*	*p*
Station	1	14.102	0.001
Plot	2	6.047	0.052
Station x Plot	2	9.609	0.003
Within + Residual	27	24.704	

[a]The repeated measures factor (time) was significant in all cases (p<0.05).

Table 2.6. Overall mean growth in terms of height (RGR_H) and numbers of internode produced of the three mangrove species at Pak Kwang and Ai Ho canals. Presented are means ± SE.

species	RGR_H (mm cm^{-1} month^{-1})		internode (internode month^{-1})	
	Pak Kwang	Ai Ho	Pak Kwang	Ai Ho
Avicennia alba	0.436 ± 0.017	0.777 ± 0.033	0.649 ± 0.028	1.011 ± 0.041
Rhizophora mucronata	0.720 ± 0.036	0.743 ± 0.035	0.568 ± 0.013	0.579 ± 0.013
Sonneratia caseolaris	1.395 ± 0.108	2.158 ± 0.139	1.855 ± 0.096	2.918 ± 0.138

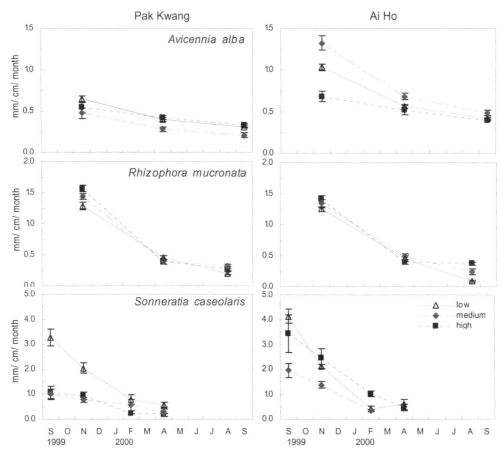

Figure 2.6. Relative growth rate expressed in height (mm cm^{-1} month^{-1}) of *Avicennia*, *Rhizophora* and *Sonneratia* seedlings at three plots of different neighbouring tree density at Pak Kwang and Ai Ho canals. Presented are means ± SE. The graphs of each species have different scale on vertical axis.

Subsequently, the data of each species were analyzed separately in order to examine the effect of the larger (station) and smaller (plot) scales of exposure on each species individually. The significant difference in RGR$_H$ was detected between the two stations for all species (p<0.05, Table 2.5). The factor 'plots' was not significant for *Avicennia* and barely significant for *Sonneratia* (p=0.052, Table 2.4) but highly significant for *Rhizophora* (P<0.001, Table 2.5). RGR$_H$ of *Avicennia* and *Sonneratia* seedlings at Ai Ho Canal was substantially higher than at Pak Kwang Canal (Table 2.6). Similar to seedling survival, *Rhizophora* seedlings exhibited a significantly higher RGR$_H$ in the high-density plot (0.753 ± 0.058 mm cm^{-1} month^{-1}) than in the low density plot (0.715 ± 0.058 mm cm^{-1} month^{-1}).

Number of internodes produced

Monthly internode production was analyzed in the same way as RGR$_H$. We found that the effect of time (repeated measures) and its interaction with stations were not significant while its interaction with species and with plots were significant (p<0.05). The number of internodes produced per month was significantly different among species, among plots and between the two stations (P<0.001, Table 2.7). The interaction between stations and species as well as stations and plots were high significant (P<0.001, Table 2.7) while the interaction between plots and species was just significant (P<0.05, Table 2.7). The same as RGR_H, a significant difference in monthly internode production between the two stations was detected for *Avicennia* and *Sonneratia* (P<0.05, Table 2.7) but not for *Rhizophora*. In addition, the factor plots was highly significant for *Rhizophora* (P<0.001, Table 2.7) but was just significant for *Sonneratia* and *Avicennia* (P<0.05, Table 2.7). The interaction between stations and plots was significant for all three species (P<0.05). In accordance with height increment, the *Sonneratia* seedlings had the highest monthly internode production while *Rhizophora* had the lowest, and the internode production of *Avicennia* and *Sonneratia* seedlings at Ai Ho Canal were substantially higher than at Pak Kwang Canal (Table 2.6, Figure 2.7). *Rhizophora* seedlings had a higher internode production in the lower neighbouring plant density plot (0.658 ± 0.017, 0.539 ± 0.014 and 0.532 ± 0.015 internode month^{-1} at the plots of low, medium and high density, respectively).

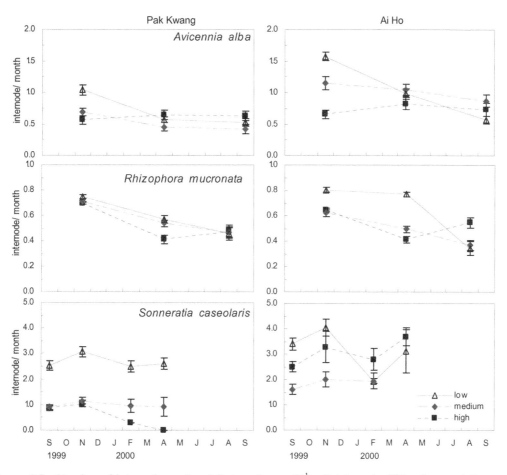

Figure 2.7. Number of internode produced (internode month^{-1}) of *Avicennia*, *Rhizophora* and *Sonneratia* seedlings at three plots of different neighbouring tree density at Pak Kwang and Ai Ho Canal. Presented are means ± SE. The graphs of each species have different scale on vertical axis.

Table 2.7. Result of repeated measures[a] GLM examining the effects species, stations and neighbouring plant densities (plots) on number of internodes produced per month for all three species and each species separately. Presented are the degree of freedom (*df*), sums of squares (*SS*) and the level of significance (*p*).

all species			
factors	*df*	*SS*	*p*
Species	2	84.406	0.000
Station	1	25.557	0.000
Plot	2	16.958	0.000
Station x Plot	2	12.248	0.000
Station x Species	2	22.241	0.000
Plot x Species	4	13.667	0.014
Station x Plot x Species	4	18.259	0.000
Within + Residual	222	62.010	

Avicennia			
factors	*df*	*SS*	*p*
Station	1	4.242	0.000
Plot	2	0.796	0.016
Station x Plot	2	0.598	0.043
Within + Residual	64	5.769	

Rhizophora			
factors	*df*	*SS*	*p*
Station	1	0.005	0.604
Plot	2	0.625	0.000
Station x Plot	2	0.361	0.000
Within + Residual	131	2.518	

Sonneratia			
factors	*df*	*SS*	*p*
Station	1	21.420	0.003
Plot	2	15.591	0.033
Station x Plot	2	22.971	0.002
Within + Residual	27	54.503	

[a]The repeated measures factor (time) was not significant (p=0.121) as well as its interaction with stations (p=0.288), while its interaction with species and its interaction with density were significant (p<0.05).

Discussion

As confirmed by our results, water movement varied between the two stations and among plots of different existing tree density, with most of the variation being seasonal. Water movement was higher in the monsoon season of October - January than in the other months of the year. The variation of shelter at the smaller (plot) scale was considerable higher than at the larger (station) scale. Shelter increased gradually from the mangrove forest edge to the interior, similar to patterns found by others (Leonard & Luther 1995). We found that water movement at the bottom of this mangrove-fringed bay was relatively low as the deployed clod cards dissolved less than 5% of their initial weight per day only. The diffusion factor (DF) values calculated from our data ranged between 2 and 9, which is considerably less than the ranges reported for tropical intertidal seagrass beds (10-15; Erftemeijer & Herman 1994) and a coral reef flat and lagoon (5-25; Jokiel & Morrissey

1993). This is not unexpected, given the fine texture of sediment of Pak Phanang Bay (Kamp-Nielsen et al. 2002), indicative of low sediment erosion rates, and low velocities reported for mangrove systems elsewhere (Wolanski 1992).

Apparently, the initial size of the seedlings was large enough to prevent them from being attacked by crabs. Therefore, none of them were damaged by this sort of predator. In addition, only few crabs were observed in the area. Seedling survival differed significantly among the three species and with increasing tree density among plots. Seedling mortality was spread equally over the year in *Avicennia*, which was not the case in the other two species. In *Sonneratia*, two periods of increased mortality were observed, which coincided with periods of heavy rainfall. Its smaller seedlings (cf. Table 1.2, Chapter 1) may be more easily dislodged during heavy rains concomitant strong currents than those of the other species. After one year, the mortality rate in the low plant density plots was highest in *Rhizophora* but lowest in *Avicennia* and *Sonneratia* (Table 2.4). The observed maximum survival in the most exposed and open plots of both *Avicennia* and *Sonneratia* is in agreement with their role as pioneer species (Tomlinson 1986; Clough 1982; Panapitukkul et al. 1998). Our observation suggested that the main cause of high mortality of *Rhizophora* in the most exposed plots was uprooting by waves or drifting objects colliding against the lengthy, rigid seedlings. The higher mortality in the exposed, mangrove front area may be a factor explaining the later appearance along the successional series by this species (Tomlinson 1986).

Seedling growth varied strongly among species. Growth of *Avicennia* and *Sonneratia* differed significantly between the inner and outer sites of the bay. *Avicennia* had a higher seedling growth at the exposed site of Ai Ho, near natural stands, than at the inner site of Pak Kwang (Figure 2.6; 2.7 and Table 2.6). Growth of *Sonneratia* followed the same pattern but its survival was higher at Pak Kwang, near natural stands, than at Ai Ho. Growth of *Rhizophora* was not significantly different between the two bay sites but RGR_H increased with increasing neighbouring vegetation densities. The seedlings of *Sonneratia* had the highest growth both in terms of height increase and production of internodes along the main stem. We found the annual number of internodes produced for *Sonneratia*, *Avicennia* and *Rhizophora* to be 30.3 ± 1.7, 13.2 ± 0.3 and 6.5 ± 0.2 respectively, which did not differ substantially from those of naturally established seedlings (28.8 ± 2.1 and 17.6 ± 0.8 for *Sonneratia* and *Avicennia*; Panapitukkul et al. 1998, Duarte et al. 1999; 6.1 ± 0.3 for *Rhizophora*; Thampanya et al. unpublished, Duarte et al. 1999). The observed decline in relative seedling growth with age is a common phenomenon in terrestrial plants ("ontogenetic drift", Van Andel & Jager 1981).

Our results confirm the higher survival of *Avicennia alba* and *Sonneratia caseolaris* at the outer, colonizing front of the mangrove forest, and provide an explanation for their role as pioneer species in the Pak Phanang Bay mangrove area (Panapitukkul et al. 1998). These species appear to be affected negatively by the density of adult plants. These effects may be derived from the shading associated to increasing tree density. In contrast, *Rhizophora* shows an opposite pattern, with the best seedling performance associated to denser stands, in agreement with their successional role as climax species in natural SE Asian mangrove forests. The first two species are also capable of establishing themselves on the open mudflat, so that mangrove rehabilitation is probably not needed if natural colonization by these species is not prevented. In addition, the cost for replanting mangrove using *Rhizophora* species is relatively high: in Thailand, this was estimated at 500 US$ per ha (Plathong 1998). However, intensive human exploitation activities (such as fishing by push net and trawler) and traffic across the mudflat by local fishing boats presently represent major sources of disturbance, dislodging numerous mangrove seedlings. The creation of temporary reserved areas excluding these activities may be sufficient to enable natural colonization, since successful seedlings generally have established well within a few years. If the replanting of economical species such as *Rhizophora* is required, areas with some existing plants are suggested as the most suitable for planting.

In summary, our results demonstrate contrasting effects of exposure and neighbouring plant density on the performance of the seedlings of the three most important species in the mangrove forest studied. These effects are consistent with the role of the species in the successional series, suggesting exposure and density to be important determinants of mangrove species succession. The results obtained also provide a basis for the selection of suitable target species for afforestation programmes in different mangrove areas.

Acknowledgements

The authors are grateful to the Coastal Resources Institute and the Faculty of Natural Resources of the Prince of Songkla University, in general, and to Dr. Somsak Boromthanarat, in particular, for giving useful suggestions and providing all research facilities. We thank Nawee Noon-anand, Chamnan Worachina, Phollawat Saijung and the fishermen at Pak Phanang West village for their friendly assistance in field data collection. We are also grateful to Prof. Patrick Denny for critically reading the manuscript. This study was funded by the Netherlands Foundation for the Advancement of Tropical Research: WOTRO project WB 84-412.

References

Clarke, P.J. and Allaway. W.G. 1993. The regeneration niche of the grey mangrove (*Avicennia marina*): effects of salinity, light and sediment factors on establishment, growth and survival in the field. Oecologia 93: 548-556

Clarke, P.J. 1995. The population dynamics of the mangrove *Avicennia marina*; demographic synthesis and predictive modelling. Hydrobiologia 295: 83-88.

Clough, B.F. 1982. Mangrove ecosystems in Australia. Australian Institute of Marine Science and Australian National University Press, Canberra.

Delgado, P., Jiménez, J.A. and Justice, D. 1999. Population dynamics of mangrove *Avicennia bicolor* on the Pacific coast of Costa Rica. Wetland Ecology and Management 7: 113-120.

Doty, M.S. 1971. Measurement of water movement in reference to benthic algal growth. Botanica Marina 14: 32-35.

Duarte, C.M., Thampanya, U., Terrados, J., Geertz-Hansen, O. and Fortes, M.D. 1999. The determination of the age and growth of SE Asian mangrove seedlings from internodal counts. Mangrove and Salt Marsh 3:251-257.

Erftemeijer, P.L.A. and Herman, P.M.J. 1994. Seasonal changes in environmental variables, biomass, production and nutrient contents in two contrasting tropical intertidal seagrass beds in South Sulawesi, Indonesia. Oecologia 99:45-59.

Field, C.D. 1995. Journey Amongst Mangroves. International Society for Mangrove Ecosystems, Okinawa, Japan.

Havanond, S. 1995. Re-afforestation of mangrove forests in Thailand. In: Khemnark C (ed), Ecology and management of mangrove restoration and regeneration in East and Southeast Asia. Proceeding of the ECOTONE IV. Kasetsart University, Bangkok, p 203-216

Hunt, R. 1982. Plant growth curves: the functional approach to plant growth analysis. Edward Arnold Pub Limited, London.

Hutchings, P. and Saenger, P. 1987. Ecology of mangroves. University of Queensland Press, St Lucia.

JICA. 1987. Basic design study report on the project for constructing the Nakhon Si Thammarat fishing port. Japan International Cooperation Agency (JICA).

Jokiel. P.L. and Morrissey, J.I. 1993. Water motion on coral reefs; evaluation of the 'clod card' technique. Marine Ecology Progress Series 93: 175-181.

Kamp-Nielsen, L., Vermaat, J.E., Wesseling. I., Borum, J., and Geertz-Hansen, O. 2002. Sediment properties along gradients of siltation in South-East Asia. Estuarine, Coastal and Self Science 54: 127-137.

Lee, S.K., Tan, W.H. and Havanond, S. 1996. Regeneration and colonisation of mangrove on clay-filled reclaimed land in Singapore. Hydrobiologia 319: 23-35.

Leonard, L.A. and Luther, M.E. 1995. Flow hydrodynamics in tidal marsh canopies. Limnological Oceanography 40: 1474-1484.

Panapitukkul, N., Duarte, C.M., Thampanya, U., Terrados, J., Keowvongsri, P., Geertz-Hansen, O., Srichai, N. and Boromthanarat, S. 1998. Mangrove colonization: Mangrove progression over the growing Pak Phanang (SE Thailand) mud flat. Estuarine, Coastal and Shelf Science 47:51-56.

Petticrew, E.L. and Kalff, J. 1991. Calibration of a gypsum source for freshwater flow measurements. Canadian Journal of Fisheries and Aquatic Sciences 48: 1244-1249.

Plathong, J. 1998. Status of mangrove forest in southern Thailand. Wetlands International – Thailand Programme/PSU, Publication No.5, 128 pp.

Porter, E.T., Sanford, L.P. and Suttles, S.E. 2000. Gypsum dissolution is not a universal integrator of 'water motion'. Limnological Oceanography 45: 145-158.

Potvin, C.P., Lechowicz, M.J. and Tardif, S. 1990. The statistical analysis of ecophysiological response curves obtained from experiments involving repeated measures. Ecology 71: 1389-1400.

Silvertown, J.W. 1982. Introduction to plant population ecology. Longman Group Limited, New York.

Thampanya, U. and Noon-anand, N. 1999. Abundance and species richness of benthic fauna in different mangrove forest ages: case study of Songkhla Lake, Thailand. Prince of Songkla University – CORIN internal report.

Tomlinson, P.B. 1986. The botany of mangroves. Cambridge University Press, Cambridge.

Tomlinson, P.B. 2000. Systematic and functional anatomy of seedlings in mangrove Rhizophoraceae: Vivipary explained? Botanical Journal of the Linnean Society 134: 215-231.

Van Andel, J. and Jager, J.C. 1981. Analysis of growth and nutrition of six plant species of woodland clearings. Journal of Ecology 69: 871-882.

Wolanski, E. 1992. Hydrodynamics of mangrove swamps and their coastal waters. Hydrobiologia 247: 141-161.

Chapter 3

The effect of increasing sediment accretion on the seedlings of three common Thai mangrove species

Udomluck Thampanya, Jan E. Vermaat and Jorge Terrados

Abstract

Three to four months old seedlings of three common Thai mangrove species (*Avicennia officinalis* L., *Rhizophora mucronata* Lamk and *Sonneratia caseolaris* (L.) Engler) were experimentally buried using six sediment accretion levels (0, 4, 8, 16, 24 and 32 cm) in a randomized block design for approximately one year. *Avicennia* was fivefold more sensitive to burial than *Sonneratia* and the seedlings of the latter species exhibited the lowest mortality as well as the highest growth rate. The numbers of surviving seedlings of these two species were highly affected by burial ($p<0.001$) and their survival decreased with increasing sediment accretion. The seedlings receiving 32 cm of sediment had the highest mortality (100% in *Avicennia*, 70% in *Rhizophora* and 40% in *Sonneratia*). Survival of *Rhizophora*, however, was not significantly different amongst burial treatments ($p=0.23$). Natural mortality in the control seedlings was substantial in *Avicennia* and *Rhizophora* (10% and 40%, respectively). The burial had significant effects on seedling height only in *Avicennia* and *Sonneratia* ($p<0.05$). The relative growth rate in terms of height was lowest in the 32 cm treatment in both species: 0.30 ± 0.19 and 1.20 ± 0.11 mm cm^{-1} month^{-1}, respectively, compared to 1.15 ± 0.15 and 1.28 ± 0.09 mm cm^{-1} month^{-1} in the controls. Annual internode production was not significantly affected by burial in any species. Although the seedling survival of *Rhizophora* was not significantly affected by different sediment levels, the overall survival of this species was much lower than that of *Sonneratia*. The results reveal that *Sonneratia* will be better suited for colonizing or planting in areas where abrupt high sedimentation is possible.

Based on Thampanya, U., Vermaat, J.E. Terrados, J. 2002. The effect of increasing sediment accretion on the seedlings of three common Thai mangrove species. Aquatic Botany 74: 315-325.

Introduction

Mangroves are common inhabitants of tropical sheltered mudflats, lagoons and deltas. The adult trees of many mangrove species show distinctive aerial roots with a well-developed system of interconnected lacunae and lenticels that store and transport oxygen inside the plant (Tomlinson, 1986; Hutchings & Saenger 1987; Ashford & Allaway 1995). Aerial roots are generally recognized as an adaptation to cope with the oxygen deficiency and reduced conditions that render a number of soluble phytotoxins such as Fe^{2+}, Mn^{2+}, and H_2S (McKee 1993; Youssef & Saenger 1996; 1998). In young seedlings, whose aerial roots have not yet been developed, lenticels are conspicuous on hypocotyl, stem and internodes such as in *Rhizophora* and *Avicennia* (Youssef & Saenger 1996; Ashford & Allaway 1995). During high tide, these lenticels are covered with water preventing gas exchange through their aerial vents. Tidal submergence may well cause short-term oxygen shortage in the roots, particularly during night time, but subsequent low tide emergence will re-establish atmospheric contact and resume root aerobic respiration. The venting role of the lenticels is, however, impaired if they are covered by sediment. Tropical rivers carry massive amounts of sediments during the rainy season (Milliman & Meade 1983) that are discharged into the coastal area, frequently as sudden high-sedimentation events. Such events can cause extensive burial of the mangrove aerial roots, inhibit root aeration, and consequently, lead to widespread mortality (Hutchings & Saenger, 1987). A number of accounts on impacts of sedimentation on the mangroves have been compiled recently by Ellison (1998). However, relevant quantitative data on SE Asian mangrove species and particularly on seedlings which might be more sensitive to burial than adult trees, are rare (Terrados et al. 1997).

The aim of this study is to examine the effects of experimental sediment burial on seedlings of three common SE Asian mangrove species: *Avicennia officinalis* L., *Sonneratia caseolaris* (L.) Engler and *Rhizophora mucronata* Lamk. The first two species have been considered as pioneer species colonising newly accreting mud flats (Tomlinson 1986; Lee et al. 1996; Panapitukkul et al. 1998) and the last is a dominant, late successional species widely used for rehabilitation programs (Havanond 1995; Field 1995). Knowledge of species-specific sensitivities to sedimentation will provide a better understanding of natural processes of space occupation by mangroves (i.e. colonization on new habitats, recovery after perturbation) which is useful for planning of more successful mangrove rehabilitation schemes (Havanond 1995; Elster 2000).

Material and Methods

Study area

The study was conducted at the edge of the mangrove forest in Songkhla Lake, Songkhla province, Southern Thailand (Figure 3.1). Songkhla Lake comprises of three connected lakes with the total area of 987 km^2. The lakes are shallow (1-2 m depth) except in the navigation channels (4-7 m). Tides are mixed-diurnal and the tidal level varies between 0.1-0.6 m. The area encounters two climatic periods; a short dry period (Feb-May) and a long rainy period with the SW monsoon (May-Sep) and the NE monsoon (Oct-Jan). Spatial salinity ranges between 20-28 ppt during the dry season and 0-14 ppt during the rainy season. Annual mean temperature is 28°C and average rainfall is 2150 mm. Many canals carrying suspended sediments from various sources discharge into the Lake. During the rainy season, huge amounts of sediment are deposited and accumulate in the area (496,000 m^3 year^{-1}; Harbour Dept 1996). Mangrove fringes are present discontinuously along the east bank of the Lake. *Sonneratia caseolaris* and *Nypa fruiticans* are abundant in the upper part of the Lake whilst some *Avicennia officinalis* mixed with *Rhizophora apiculata* and *Rhizophora mucronata* occupy the lower part.

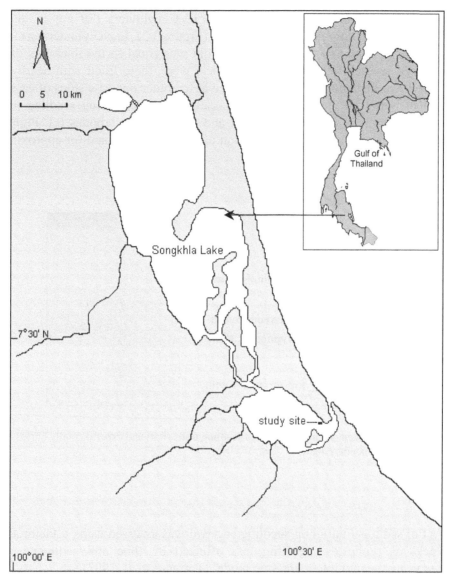

Figure 3.1. Map of Songkhla Lake showing location of the study site.

Burial experiment

The experiment was set up at the mangrove forest edge near the Hua Khao community participatory rehabilitation area (7° 11′N 100° 33′E), six km from the mouth of the Lake (Figure 3.1). Soil in the area is composed of 22-30% clay and 28-31% silt. Between 3-4 months old seedlings of *Avicennia officinalis*, *Rhizophora mucronata* and *Sonneratia caseolaris*, with average heights of 46, 77 and 36 cm, respectively, were transplanted with

1.5 m x 1.5 m spacing in May 1999. *Avicennia* and *Sonneratia* seedlings were obtained by shovelling natural seedlings available in the area while *Rhizophora* seedlings were obtained from the community's nursery. A randomized block design with five blocks was applied for 60 seedlings of each species. After transplanting, the seedlings were allowed to recover for three weeks before being buried with sediment. In each block, six sediment burial levels (0, 4, 8, 16, 24, 32 cm) with two replicates were assigned randomly to the seedlings. Sediment burial was realized with PVC cylinders (30 cm in diameter) of different height encircling individual seedlings (Figure 3.2). Each cylinder was inserted 10 cm into the substratum and the remaining height was equal to the thickness of sediment burial assigned to each individual. All cylinders were, then, filled with nearby available sediment and two bamboo sticks were inserted at the outer opposite sides of each cylinder. The effect of burial on the seedlings was assessed by quantifying seedling growth and mortality. Measurement of seedling height and number of internodes (cf. Figure 3.2) on the main stem was conducted at the initiation of the experiment and for approximately one year with three-month intervals.

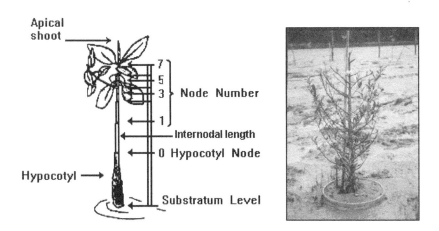

Figure 3.2. Burial experiment set up: Quantified seedling traits (left) and experimentally buried *Sonneratia* seedling (4 cm) after 6 months (right).

Data analysis

The effect of sediment burial on seedling survival was assessed using a factorial ANOVA with blocks as replicates. Seedling loss affected by time and sediment burial was quantified using the multiple regression model (Terrados et al. 1997) as:

$$\text{survivors} = a + b_1(\text{time}) + b_2(\text{burial} \times \text{time})$$

where:

   b_1 is a daily loss rate with time

and

   b_2 is an increase in daily loss rate per cm of sediment burial, i.e. the interaction of burial depth with time.

Because of a slight difference in initial seedling sizes, data on seedling height were translated into relative growth rate (Hunt 1982). An ANOVA with repeated measures design (Potvin et al. 1990) was applied for the examination of burial effect on seedling growth both in terms of relative growth rate and internode production since the same seedlings were monitored on consecutive samplings.

Results

Mortality was limited in the first 1-2 months, but thereafter the three species displayed different survival patterns as a function of burial (Figure 3.3): *Sonneratia* exhibited a lower mortality than *Avicennia* and *Rhizophora*. The number of surviving seedlings differed significantly amongst sediment burial levels in *Avicennia* and *Sonneratia* (p=0.000; Table 3.1, Figure 3.3). The survival curves of *Avicennia* seedlings receiving 4 cm of sediment and the controls (0 cm) were similar and their mortality after one year was low (10%; Figure 3.3). The seedlings receiving 8 and 16 cm treatments followed the same pattern but suffered a somewhat higher mortality (20% and 30%, respectively, after one year). Survival of *Avicennia* was limited at higher burial levels: only 10% remained alive under 24 cm, whereas all had died under 32 cm already after 8 months (Figure 3.3). In *Rhizophora*, the number of surviving seedlings declined evenly in all treatments as the experiment progressed. In the controls of this species, the final survival was 60%, whereas under 4 and 8 cm of sediment this was 40% and only 30% under 32 cm (Figure 3.3). The ANOVA analysis, however, failed to detect significant differences between burial levels (p = 0.230; Table 3.1). The seedlings of *Sonneratia* survived better than those of the other two species with no mortality in the controls and the seedlings buried with 4 and 8 cm of sediment suffered only 10% mortality. Furthermore, the survival at the end of the experiment of the seedlings under 32 cm burial level was 60%, which is considerably higher than that shown by *Avicennia* and *Rhizophora* under the same burial level (Figure 3.4).

Table 3.1. Threeway ANOVA examining the effects of sediment burial on seedling survival after one year of *Avicennia officinalis*, *Rhizophora mucronata*.and *Sonneratia caseolaris*. Presented are the degrees of freedom (df), sums of squares (SS) and level of significance (p).

factors	*Avicennia*			*Rhizophora*			*Sonneratia*		
	df	*SS*	*p*	*df*	*SS*	*p*	*df*	*SS*	*p*
Treatment	5	18.8	0.000	5	3.0	0.230	5	2.4	0.000
Block	4	1.4	0.490	4	5.6	0.013	4	0.6	0.196
Treatment x Block	20	5.6	0.840	20	8.0	0.527	20	4.6	0.001
Residual	120	50.0		120	50.8		120	11.2	
Total	149	75.8		149	67.4		149	18.8	

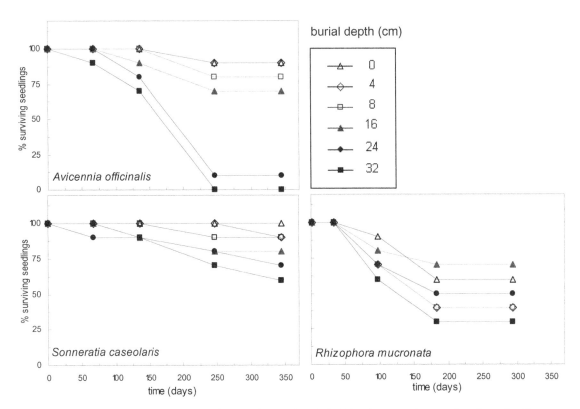

Figure 3.3. Seedling survival (%) over the experimental period of the three species as a function of sediment burial. Legend gives burial depth (cm).

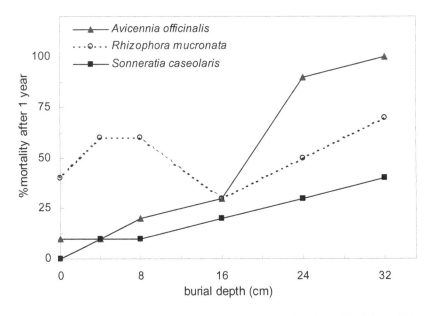

Figure 3.4. Seedling mortality (%) at the end of the experiment as a function of burial depth. Regressions were significantly in *Avicennia* and *Sonneratia* but not in *Rhizophora*.

The loss of seedlings during the experiment was modelled for each species using the following non-linear equations:

Avicennia:

survivors = 10.430 + 0.000 (± 0.002) time – 0.001 (± 0.0001) burial x time (r^2=0.87, p<0.0001)

Rhizophora:

survivors = 10.207 – 0.015 (± 0.002) time – 0.0001 (± 0.0001) burial x time (r^2=0.78, p<0.0001)

Sonneratia:

survivors = 10.004 – 0.001 (± 0.001) time – 0.0002 (± 0.0001) burial x time (r^2=0.51, p<0.0001)

These regressions indicate that the factor time singly was a negligible source of seedling loss for *Avicennia* (b_1=0, p=0.998: regression table see Annex 3.1) and *Sonneratia* (b_1= -0.001, p=0.270) but was a highly significant source of seedling loss for *Rhizophora* (b_1= -0.015, p=0.000). In contrast, sediment burial significantly increased the loss of seedlings in *Avicennia* ($b_{2 =}$ -0.001, p=0.000) and *Sonneratia* (b_2 = -0.0002, p=0.002). The number of surviving seedlings of these two species decreased at a rate of 0.001 and 0.0002 seedlings per day per cm of sediment burial, respectively. The absence of the burial effect on *Rhizophora* (Table 3.1) was confirmed by the multiple regression: the burial x time factor was not significant (b_2 = -0.0001, p=0.128). *Rhizophora* seedlings died naturally at a rate of 0.015 seedlings per day (1 dead seedling in 67 days).

Temporal differences in seedling growth were significant, both in terms of internode production and height increase, and especially for *Avicennia* and *Sonneratia* (p<0.0001; Table 3.2). The burial had a significant effect on the seeding relative growth rate (RGR_H) in *Avicennia* (p=0.008) and *Sonneratia* (p=0.026) but not in *Rhizophora* (p=0.986; Table 3.2). The RGR_H of *Avicennia* was highest in the control seedlings (1.15 ± 0.15 mm cm^{-1} month^{-1}; Figure 3.4) and declined gradually with increasing sediment levels. The seedlings of this species buried with the highest sediment depth (32 cm) had the lowest growth rate (0.30±0.19 mm cm^{-1} month^{-1}) and eventually died. In *Rhizophora*, the seedling RGR_H did not vary substantially amongst sediment levels and the seedlings receiving 4 cm of sediment had the highest RGR_H (0.22 ± 0.01 mm cm^{-1} month^{-1}; Figure 3.5) while the lowest RGR_H was observed in the 32 cm treatment. *Sonneratia* showed a higher growth rate than the other two species as well as had the lowest RGR_H at the deepest burial. The sediment burial seemed to affect the internode production in the first stage of the experiment. A significant effect on overall internode production, however, was not detected in any of the three species studied (Table 3.3). In addition, the *Sonneratia* seedlings produced a higher number of internodes (between 3.26 ± 0.37 to 3.68 ± 0.46 internodes month^{-1}; Figure 3.5) than those of *Avicennia* and *Rhizophora*, similar to the RGR_H.

Table 3.2. Result of ANOVA with repeated measures[a] examining the effects of sediment burial on height increase (RGR_H). Presented are the degrees of freedom (*df*), sums of squares (*SS*) and level of significance (*p*).

factors	Avicennia			Rhizophora			Sonneratia		
	df	*SS*	*p*	*df*	*SS*	*p*	*df*	*SS*	*p*
Treatment	4	3.50	0.008	5	0.03	0.986	5	2.44	0.027
Block	4	3.23	0.011	4	0.04	0.931	4	1.87	0.039
Treatment x Block	11	15.79	0.000	13	0.26	0.899	20	2.29	0.722
Within + Residual	14	2.33		6	0.27		19	2.84	

[a]The repeated measures factor (time) was significant in all three species (tests of within subject contrast, $p<0.001$ for *Avicennia officinalis* and *Sonneratia caseolaris*, $p<0.05$ for *Rhizophora mucronata*).

Table 3.3. Result of ANOVA with repeated measures[a] examining the effects of sediment burial on internode production. Presented are the degrees of freedom (*df*), sums of squares (*SS*) and level of significance (*p*).

factors	Avicennia			Rhizophora			Sonneratia		
	df	*SS*	*p*	*df*	*SS*	*p*	*df*	*SS*	*p*
Treatment	4	0.90	0.616	5	0.13	0.916	5	0.36	0.889
Block	4	1.73	0.316	4	0.20	0.730	4	0.74	0.523
Treatment x Block	11	2.89	0.645	13	0.14	0.999	20	4.02	0.591
Within + Residual	14	4.64		6	0.58		19	4.24	

[a]The repeated measures factor (time) was significant in all three species (tests of within subject contrast, $p\leq0.001$). Also the interaction of treatment and time was significant for *Avicennia* and *Sonneratia*. Two way ANOVA (not shown) confirmed that treatment was only significant in the first two periods.

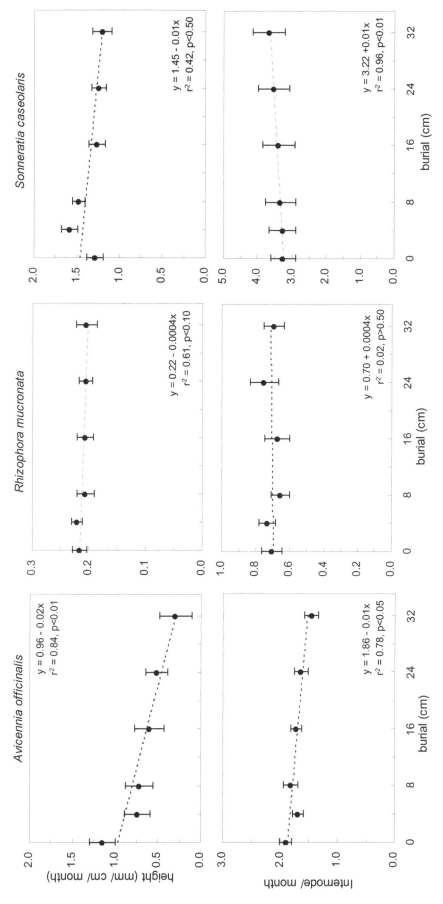

Figure 3.5. Average relative height increase (RGR$_H$) and number of internodes produced over the experimental course versus burial depth. Overall repeated measures are presented in Table 3.3.

Discussion

The three mangrove species tested in this experiment were found to differ substantially in their sensitivities to sediment burial. *Avicennia* survival decreased non-linearly with the level of burial resulting in severe mortality of the seedlings when buried with 32 cm of sediment. *Sonneratia* seedlings showed a similar pattern but the burial effect was smaller. In contrast the *Rhizophora* seedlings were not significantly affected by the burial levels and the seedling loss should be attributed to other factors not accounted for in this study The increase in sensitivity to burial above 16 cm of the *Avicennia* seedlings is in agreement with the observation in Ellison's review (1998) which suggests that burial beyond 10 cm can cause mortality. The seedlings of *Sonneratia* in particular, survived remarkably well, grew faster than those of the other species and produced the highest number of internodes (Figure 3.5). Additionally, we observed already the first pneumatophores in this species after six months and these extended outside the experimental burial cylinder. The first pneumatophores in *Avicennia* are reportedly observed in the second year only (Ashford & Allaway 1995; Skelton & Allaway 1996). The development of pneumatophores as well as the rapid elongation of the stem probably explains why *Sonneratia* coped so well with the burial.

Contrary to the finding of Terrados et al. (1997) for *R. apiculata*, sediment burial was found to have no significant effect for *R. mucronata* here. This may be due to the larger size of the propagule of the latter species (Table 1.2: 50-70 cm in *R. mucronata* vs. 20-30 cm for that of *R. apiculata*; Tomlinson 1986). The remaining non-buried part of the seedling probably contains a number of lenticels (Youssef & Saenger 1996) allowing a better aeration for *R. mucronata* and hence less sensitivity to burial. Additionally, the finer sediment used by Terrados et al. (1997) in Pak Phanang (40% clay content vs. 25% in Songkhla: Thampanya unp. data.) may have created a more adverse rooting environment causing seedling mortality although mangroves are known to establish well on very fine sediments (Ukpong 1992). At least, this contrast justifies the comparative inclusion of several species in the present experiment.

Similar to survival, seedling growth was affected differently by sediment burial in the three tested species, whereas temporal differences (as repeated-measures) were always significant. The height increase of *Sonneratia* was the highest: burial had no significant effect on the growth of *Rhizophora*. The linear regressions of the overall internode production for *Avicennia* and *Sonneratia* were significant (Figure 3.5). This was possibly due to a significant difference in internode production at the early stage of the experiment. The ANOVA repeated measures, however, failed to detect significant effects of sediment burial on internode production in any species. This supports the findings of Duke and Pinzon (1992) and Duarte et al. (1998) that the annual number of internodes produced by a mangrove apex is a rather conservative and species-specific trait. Internodal lengths, however, vary seasonally as in e.g. seagrasses (Duarte et al. 1994) and may be highly sensitive to environmental changes (as e.g. burial). Apparent differences in annual numbers of internodes produced within a species among sites or years were not significant (Thampanya et al. 2002, Chapter 2), but do point at variability in this trait.

We have demonstrated in this study that sensitivity to burial varies substantially among species. Different taxa probably cope differently with the stress due to burial (reducing conditions and less oxygen provided to root system). The size of a taller seedling of *Rhizophora mucronata* and the rapid development of pneumatophores in a faster growing

Sonneratia appear to be useful strategies. Early colonizing species may encounter strong water currents as well as prolonged water submergence, necessitating a rapid development of the root system and stem elongation as shown here for *Sonneratia*. In rehabilitation schemes, *Sonneratia* is probably an efficient colonizer for areas with prolonged inundation or high sedimentation. *Avicennia* is probably suitable in areas exposed to high water turbulence (Thampanya et al. 2002, Chapter 2) as long as sediment deposition does not exceed 16 cm. *Rhizophora*, although not significantly affected by sediment burial, appears to be less suitable for plantation in areas exposed to wind and waves (Chapter 2). Pilot field experiments prior to large-scale plantation schemes are advisable for reforestation programs using this economical important species.

Acknowledgements

The authors wish to thank Dr. Somsak Boromthanarat and Dr. Carlos M. Duarte for useful advice. We also thank the Coastal Resources Institute, Prince of Songkla University for providing all research facilities and Chamnan Worachina, Phollawat Saijung for assistance in the field. This research was funded by the Netherlands Foundation for the Advancement of the Tropical Research (WOTRO grant WB84-412).

References

Ashford, A.E. and Allaway, W.G. 1995. There is a continuum of gas space in young plants of *Avicennia marina*. Hydrobiologia 295: 5-11.

Chapman, V.J. 1976. Mangrove vegetation. J. Cramer, Vaduz.

Duarte C.M., Marbá, N. and Agawin, N. 1994. Reconstruction of seagrass dynamic: age determination and associated tools for the seagrass ecologist. Marine Ecology Progress Series 107: 195-209.

Duarte, C.M., Thampanya, U., Terrados, J., Geertz-Hansen, O. and Fortes, M.D. 1999. The determination of the age and growth of SE Asian mangrove seedlings from internodal counts. Mangroves and Salt Marshes 3: 251-257.

Duke, N.C. and Pinzon, Z.S. 1992. Aging *Rhizophora* seedlings from leaf scar nodes: a technique for studying recruitment and growth in mangrove forests. Biotropica 24: 173-186.

Ellison, J.C. 1998. Impacts of sediment burial on mangroves. Marine Pollution Bulletin 37: 420-426.

Elster, C. 2000. Reason for reforestation success and failure with three mangrove species in Colombia. Forest Ecology and Management 131: 201-214.

Field, C.D. 1995. Journey Amongst Mangroves. International Society for Mangrove Ecosystems, Okinawa, Japan.

Harbour department. 1996. Report on hydrological survey of Songkhla deep-sea port. Bangkok, Thailand, 35 pp.

Havanond, S. 1995. Re-afforestation of mangrove forests in Thailand. In: Khemnark, C. (ed), Ecology and Management of Mangrove Restoration and Regeneration in East and Southeast Asia. Proceeding of the ECOTONE IV. Kasetsart University, Bangkok, pp. 203-216.

Hovenden, M.J., Curran, M., Cole, M.A., Goulter, P.F.E., Skelton, N.J. and Allaway, W.G. 1995. Ventilation and respiration in roots of one-year-old seedlings of grey mangrove *Avicennia marina* (Forsk.) Vierh. Hydrobiologia 295: 23-29.

Hunt, R. 1982. Plant growth curves: the functional approach to plant growth analysis. Edward Arnold Pub Limited, London.

Hutchings, P. and Saenger, P. 1987. Ecology of Mangroves. University of Queensland Press, St Lucia.

Lee, S.K., Tan, W.H. and Havanond, S. 1996. Regeneration and colonisation of mangrove on clay-filled reclaimed land in Singapore. Hydrobiologia 319: 23-35.

McKee, K.L. 1993. Soil physicochemical patterns and mangrove species distribution – reciprocal effect? Journal of Ecology 81: 477-487.

Milliman, J.D. and Meade, R.H. 1983. World-wide delivery of river sediments to the oceans. Journal of Geology 91: 1-21.

Panapitukkul, N., Duarte, C.M., Thampanya, U., Terrados, J., Keowvongsri, P., Geertz-Hansen, O., Srichai, N. and Boromthanarat, S. 1998. Mangrove colonization: Mangrove progression over the growing Pak Phanang (SE Thailand) mud flat. Estuarine, Coastal and Shelf Science 47: 51-56.

Potvin, C.P., Lechowicz, M.J. and Tardif, S. 1990. The statistical analysis of ecophysiological response curves obtained from experiments involving repeated measures. Ecology 71: 1389-1400.

Skelton, N.J. and Allaway, W.G. 1996. Oxygen and pressure changes measured in situ during flooding in root of grey mangrove *Avicennia marina* (Forsk.) Vierh. Aquatic Botany 54: 165-175.

Terrados, J., Thampanya, U., Srichai, N., Keowvongsri, P., Geertz-Hansen, O., Boromthanarat, S., Panapitukkul, N. and Duarte, C.M. 1997. The effect of increased sediment accretion on the survival and growth of *Rhizophora apiculata* seedlings. Estuarine, Coastal and Shelf Science 45: 697-701.

Thampanya, U., Vermaat, J.E. and Duarte, C.M. 2002. Colonization success of common Thai mangrove species as a function of shelter from water movement. Marine Ecology Progress Series 237: 111-120.

Tomlinson, P.B., 1986. The botany of mangroves. Cambridge University Press, Cambridge.

Ukpong, I. 1992. The interrelationships between mangrove vegetation and soils using multiple regression analysis. Ekologia Polska, 40: 101-112.

Youssef, T. and Saenger, P. 1996. Anatomical adaptive strategies to flooding and rhizosphere oxidation in mangrove seedlings. Australian Journal of Botany 44: 297-313.

Youssef, T. and Saenger, P. 1998. Photosynthetic gas exchange and accumulation of phytotoxins in mangrove seedlings in response to soil physico-chemical characteristics associated with water logging. Tree Physiology 18: 317-324.

Annex

Annex 3.1. Results of non-linear regression models examining the effects of time and interaction of time and sediment burial on seedling survival of *Avicennia officinalis*, *Rhizophora mucronata* and *Sonneratia caseolaris*.

species	intercept	coefficient	value	SE	p
Avicennia	10.430	b_1 (time)	0.000004	0.002	0.998
		b_2 (time x burial)	-0.001	0.0001	0.000
Rhizophora	10.207	b_1 (time)	-0.015	0.002	0.000
		b_2 (time x burial)	-0.0001	0.0001	0.128
Sonneratia	10.004	b_1 (time)	-0.001	0.001	0.270
		b_2 (time x burial)	-0.0002	0.0001	0.002

Chapter 4

Diameter and height-age relationships for common SE Asian mangrove taxa

Udomluck Thampanya and Jan E. Vermaat

Abstract

Height-Diameter at Breast Height (DBH)-age regressions were satisfactorily compiled from field data and grey literature for trees up to 20 years old of the most common SE Asian mangrove taxa: *Rhizophora, Avicennia* and *Sonneratia*. The regressions of height-age or DBH-age were found to describe the observed data equally well. Linear fits were sufficient for these comparatively young trees. Intercepts of these linear regressions were generally more variable both within and among species than were slopes. Overall, slopes in DBH-age relations ranged between 0.5 and 2.3 cm y^{-1} among species, whereas those in height-age ranged between 0.6 and 1.1 m y^{-1}. Hence species differed more in wood accumulation with trunk width than in length increase. In *Rhizophora apiculata*, the slopes of both DBH-age and height-age were identical at three sites (Samut Songkram, Ranong and Pattani). However, the DBH-age slope of this species was less steep than that of *Rhizophora mucronata*: the latter species thus accumulated trunk wood more rapidly than the former. Slopes and intercepts of the three *Avicennia* species (*Avicennia alba, Avicennia marina Avicennia officinalis*) varied considerably. Among the three taxa, *Sonneratia* showed the highest slopes both in DBH-age and height-age which is in accordance with the fast growth capability of this species.

To be submitted to Forest Ecology and Management

Introduction

Mangrove forests are exploited for timber which is used for construction material, woodchip, pulpwood, charcoal and domestic firewood (Aksornkoae 1993; FAO 1982; Plathong & Sitthirach 1998; Semesi 1998; Naylor et al. 2002). In conventional forestry practice, stock assessments of available wood are achieved using easily measurable parameters, i.e. diameter at breast height (DBH), height and density, to estimate woody product (FAO 1993; RFD 2001). For relatively tall trees or dense forests, stem height is inconvenient to measure; hence various height-DBH relationship models have been developed to overcome this problem (Fang & Bailey 1998; Loewenstein et al. 2000; Peper et al. 2001; Colbert & Larsen 2002; Zeide & Vanderschaaf 2002). Nevertheless, studies on tree age estimated from these parameters are rare (Tabbush & White 1996; Hoshino et al. 2001; McDowell et al. 2002). This is also the case for mangroves where similar approaches have been applied to address issues such as economic valuation, forest structure, volume estimation, and biomass production (Sandrasegaran 1971; FAO 1994; Turner et al. 1995; Stienke et al. 1995; Tam et al. 1995; Davoe & Cole 1996; Cole et al. 1999). Furthermore, growth ring analysis for determining age of mangroves is still doubtful since no confirmed report is available as yet about the correlation between number of produced rings and stand age (Tomlinson 1986; Menezes et al. 2003; Verheyden et al. 2004). Verheyden et al. (2004), for example, suggests that both the short days and low temperatures of winter and draught stress of a dry season may form conspicuous rings in *Rhizophora mucronata* leading to two rings produced per year.

A number of SE Asian mangrove studies focus on ecosystem change over time related to environmental conditions such as ecosystem degradation, and coastal change (Kongsangchai et al. 1988; Fujimoto et al. 1996; Macintosh et al. 2002; Thampanya et al. in prep, Chapter 5). This information is crucial for coastal resource management in the region, although published quantitative data on forest structure and mangrove demography are comparatively scarce (Putz & Chan 1986; Panapitukkul et al. 1998; Duarte et al. 1999; Ha et al. 2003). The establishment of robust age-height-DBH relationships would facilitate this kind of studies. We present here a compilation of our own field data, and local forestry reports that are difficult to access, on such relationships for mangroves. Our aims were:

(1) to provide age-height-DBH regression relations for predictive purposes of the most common SE Asian mangrove taxa, and

(2) to analyze variability in these relations within and among species and sites.

Material and methods

Data collection

Growth data of three common Thai mangrove genera (*Rhizophora*, *Avicennia* and *Sonneratia*) were compiled for ages ranging from 0.5 to 20 years. These data consist of tree height (m) and diameter at breast height (DBH, cm) for trees taller than 1.5 m or older than 2-3 years. For young trees with height less than 1.5 m, only stem height has been used. Three datasets of *Rhizophora apiculata* stands, the most commonly used species for mangrove rehabilitation, were obtained from three sites: two mangrove research stations, Ranong Mangrove Research Center on the southwest coast and Pattani Mangrove Research Station in the Southeast, and the long-term research plot with a planted forest in

Samut Songkram province in the upper Gulf of Thailand (Wechakit 1987). For *Rhizophora mucronata*, *Avicennia alba*, *Avicennia officinalis*, *Avicennia marina* and *Sonneratia caseolaris*, comprehensive data of successive growth were not available. Therefore, field measurements were conducted at forests with known planting dates. These measured data were merged with data obtained from literature (Kongsangdow et al. 1988; Bamroongrugsa & Yuanlaie 1993; Havanond et al. 1994; Bamroongrugsa 1997; Kaewwongsri & Bamroongrugsa 1997a; Kaewwongsri & Bamroongrugsa 1997b; Piriyayotha and Jaicheeng 2001; Jirawattanapun et al. 2002) to accomplish our datasets for these species.

Data Analysis

We assessed the relationships of mangrove growth parameters in terms of height-age and DBH-age. Since this study considers the early phase (the first 20 years) of the mangrove life span which is reported for approximately 90-100 years or more (Verheyden et al. 2004), the data were fitted using linear regression instead of sigmoidal or non-linear forms reported elsewhere for longer-term studies of tree allometric-age relations (Sandrasegaran 1971; Pienaar & Turnbull 1973; Tabbush & White 1996; Dewar & McMurtrie 1996; Maa et al. 2002).

The height-age and DBH-age regression equations were compared to assess whether height or DBH was more closely correlated to age. The F-statistic examining the residual sums of squares (RSS) and degree of freedoms (df) of the fit was used for this comparison applying equation 1 (Lederman & Tett 1981).

$$F = \frac{RSS_{height}/df_{height}}{RSS_{DBH}/df_{DBH}} \tag{1}$$

We applied dummy variables regression to assess homogeneity of slope and intercepts among the *R. apiculata* data from three sites, between *R. apiculata* and *R.mucronata* and among the three species of *Avicennia*. Thereafter, pairwise comparisons were made. Data from Ranong and Pattani sites were combined for *R. apiculata,* then compared with data of *R. mucronata*. In the detail examination, identical in slopes and intercepts of two lines were tested. The F statistic for comparison was computed from the RSS and df of full and reduced models (Neter et al. 1996; Ott & Longnecker 2001) as in equation 2:

$$F = \frac{(RSS_{reduced} - RSS_{full})/(df_{reduced}-df_{full})}{(RSS_{full}/df_{full})} \tag{2}$$

Results

Growth parameters

Highly significant relationships of height-age and DBH-age were found in all species. The coefficient of determination (r^2) ranged from 0.94 to 0.99 for height-age and 0.81 to 0.98 for DBH-age (Table 4.1). The height-age slopes of the three sites of *R. apiculata* were similar whilst differences were noticed for the DBH-age (Table 4.1; Figure 4.1a and 4.1b). This was also evident from a graphical comparison of *R. apiculata* and *R. mucronata*

(Figure 4.1c and 4.1d). The slopes of the three *Avicennia* species were comparable (Table 4.1; Figure 4.2a and 4.2b), though significant differences existed (see below).

Table 4.1. Linear regression of height-age and DBH-age datasets of southern Thailand mangrove species. Age is independent variables. All equations are highly significant (p<0.001). [a] and [b] are Tukey comparison.

dependent variables	species	slope (SE)	intercept (SE)	r^2	n
height	*R .apiculata* - Pattani	0.77 (0.04)[a]	-0.46 (0.41)[a]	0.98	14
	- Ranong	0.83 (0.02)[a]	-0.67 (0.21)[a]	0.99	15
	- Samut Songkram	0.83 (0.06)[a]	0.04 (0.51)[b]	0.94	15
	R. apiculata (combined)	0.79 (0.02)[a]	-0.49 (0.21)[a]	0.98	29
	R. mucronata	0.78 (0.03)[a]	0.33 (0.23)[b]	0.99	9
	A. alba	0.71 (0.04)[a]	0.46 (0.31)[b]	0.97	11
	A. officinalis	0.59 (0.02)[b]	0.40 (0.12)[a]	0.99	12
	A. marina	0.69 (0.02)[a]	-0.13 (0.13)[a]	0.99	13
	S. caseolaris	1.12 (0.03)	-0.22 (0.26)	0.99	16
DBH	*R .apiculata* - Pattani	0.45 (0.03)[a]	0.48 (0.35)[b]	0.96	13
	- Ranong	0.43 (0.02)[a]	-0.34 (0.22)[a]	0.97	12
	- Samut Songkram	0.32 (0.04)[a]	1.49 (0.41)[b]	0.84	13
	R. apiculata (combined)	0.47 (0.03)[a]	-0.16 (0.33)[a]	0.92	25
	R. mucronata	0.70 (0.05)[b]	1.09 (0.44)[b]	0.97	7
	A. alba	0.75 (0.03)[a]	0.51 (0.33)[b]	0.98	9
	A. officinalis	0.83 (0.06)[ab]	-1.59 (0.57)[a]	0.96	8
	A. marina	1.04 (0.06)[b]	-2.29 (0.39)[a]	0.98	9
	S. caseolaris	2.30 (0.11)	-3.32 (0.92)	0.97	14

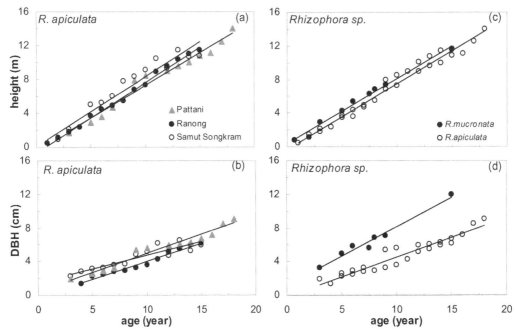

Figure 4.1. Left: Linear regression of three sites of *R. apiculata* (a) height-age, (b) DBH-age; Right: combined *R. apiculata* versus *R. mucronata* (c) height-age, (d) DBH-age.

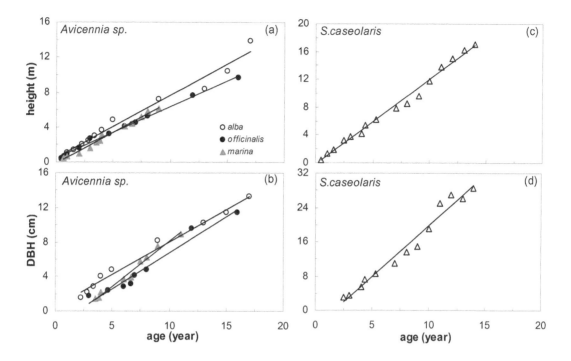

Figure 4.2. Left: Linear regression of three *Avicennia* species (a) height-age and (b) DBH-age;
Right: *Soneratia caseolaris* (c) height-age and (d) DBH-age.

Height-age and DBH-age regressions generally described the observed data equally well ($p>0.05$). In two cases, however, the height-age regression was significantly better: this was the case for *A. officinalis* and *S. caseolaris* ($p=0.044$ and $p=0.0004$, respectively cf. figure 4.2a versus 4.2b and 4.2c versus 4.2d).

Within and among species variation

The F test for homogeneity of slopes within *R. apiculata* was not significant for the height-age but it was so for DBH-age ($p=0.27$ and $p=0.02$, respectively; Annex 4.1). Among species of *Avicennia*, both height-age and DBH-age slopes were significantly different ($p=0.012$). Remarkably, significant differences were found in all coincidence tests (equal intercepts) within *R. apiculata* as well as among species of *Avicennia* ($p<0.001$).

In pairwise comparisons of slopes and intercepts (Table 4.1), slopes were only found to differ between the two species of *Rhizophora* and among species of *Avicennia*. DBH-age slope of *R. apiculata* was less steep than that of *R. mucronata* (Figure 4.1d) whilst it was steeper in *A. marina* than in *A. alba* (Figure 4.2c). For height-age regression, slopes were significant different between *A. alba* and *A. officinalis* and between *A. officinalis* and *A. marina* (Table 4.1). *A. alba* had the steepest slope followed by *A. marina* and *A. officinalis*, respectively. Intercepts differed more frequently. They were significantly different for both DBH-age and height-age, among *R. apiculata* sites, between the two *Rhizophora* and among *Avicennia* species.

Predictive models for mangrove age

The age-height and age-DBH predictive models for an estimation of mangrove stand were attained from the inverted relationships of the previous height-age and DBH-age regressions. Equations for all selected species are predicted in Table 4.2.

Table 4.2 Age-height and age-DBH models for predicting age of common Thai mangrove species.

regression	species	model	SE of y estimate
age-height	*R. apiculata* - Pattani	y = 0.19+1.33x	0.69
	- Ranong	y = 0.96+1.16x	0.20
	- Samut Songkram	y = 0.41+1.14x	1.23
	R. apiculata (combined)	y = 0.77+1.24x	0.66
	R. mucronata	y = -0.35+1.27x	0.24
	A. alba	y = -0.47+1.36x	0.95
	A. officinalis	y = -0.65+1.74x	0.74
	A. marina	y = 0.32+1.45x	0.19
	S. caseolaris	y = 0.25+0.81x	0.46
age-DBH	*R. apiculata* - Pattani	y = -0.45+2.11x	0.96
	- Ranong	y = 1.01+2.24x	0.36
	- Samut Songkram	y = -2.26+2.52x	1.75
	R. apiculata (combined)	y = 1.19+1.96x	1.28
	R. mucronata	y = -1.30+1.38x	0.46
	A. alba	y = -0.56+1.31x	0.94
	A. officinalis	y = 2.69+1.02x	0.99
	A. marina	y = 2.42+0.90x	0.14
	S. caseolaris	y = 2.60+0.42x	1.86

y=age (year); x=height (m) or DBH (cm).

Discussion

Overall, we found that both height-age and DBH-age regressions were described satisfactorily with linear regression models ($r^2 > 0.8$; $p < 0.001$) for all SE Asian mangrove taxa analysed. Generally, the height and DBH data appear to fit the observed data equally well since height-age fits were significantly better in only two cases. The standard error of our age estimates varied between 0.14 (*A. marina*) to 1.86 years (*S. caseolaris*) indicating differences in variability among datasets. For predictive ecological purposes, the models have satisfactory precision.

Our comparison of slopes and intercepts among different sites and species suggest substantial similarity within *Rhizophora*. The height-age curves were found to be similar both within *R. apiculata* as well as between *R. apiculata* and *R. mucronata*. Besides, DBH-age slopes among *R. apiculata* sites were not significant different. Intercepts, however, were found to be more variable. Possibly, the lower precipitation and higher temperatures of the upper Gulf part led to less growth in stem diameter than the humid and high rainfall of the South whilst factors influencing tree height such as light intensity and tree density were comparable. The higher DBH growth of *R. mucronata* supports qualitative observations in the field where *R. apiculata* stems are usually thinner than those of *R. mucronata* in the same forest.

Among three species of *Avicennia*, the observed variation was considerable. Both for height-age and DBH-age, *A. alba* had the highest intercept, and this may be coupled to the initial size of the seedling used in plantation (Bamroongrugsa 1997; Kaewwongsri & Bamroongrugsa 1997a). For slopes, the observed pattern is complex. *A. marina* had the highest diameter increment, whereas *A. officinalis* was intermediate and *A. alba* had the lowest slope. The latter, however, displayed the fastest height increase together with *A. marina* whereas *A. officinalis* lagged behind. A more rapid height increase of *A. marina* and *A.alba* is in agreement with their distribution as pioneers (Tomlinson 1986; Panapitukkul et al. 1998; Lee et al. 1996) whilst *A. officinalis* is observed more frequently within closed stands (Kaewwongsri & Bamroongrugsa 1997b, Satyanarayana et al. 2002).

In general, DBH-age slopes ranged between 0.5 and 2.3 cm y^{-1} whilst those of height-age ranged between 0.6 and 1.1 m y^{-1}. Hence the three species differed more in wood accumulation with trunk width than in length increase. *Sonneratia* exhibited the fastest growth among the tree taxa both in DBH and height (2.3 cm y^{-1} and 1.1 m y^{-1}, respectively). At the age of 20 year, *Sonneratia caseolaris* has 43 cm in DBH and 22 m in height. For the other species, growth patterns were comparable with height ranging from 12 m in *A. officinalis* to 16 m for *R. mucronata* and DBH ranged from 9 cm for *R. apiculata* to 19 cm for *A. marina*. This growth pattern clearly separates *Sonneratia caseolaris* from the others (Figure 4.3) and confirms the fast growth capability of this species.

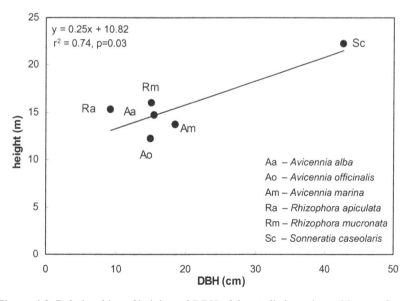

Figure 4.3. Relationships of height and DBH of the studied species at 20 year of age.

Acknowledgements

The study was funded by the Netherlands Foundation for the Advancement of Tropical Research (WOTRO: project WB 84-412). We thank Chamnan Worachina, Phollawat Saijung for field data collection and the forestry officers at Ranong Mangrove Research Center and Pattani Mangrove Research Station for providing growth data.

Reference

Aksornkoae, S. 1993. Ecology and management of mangroves. IUCN, Bangkok, 176 pp.

Bamroongrugsa, N. and Yuanlaie, P. 1993. Experimentation for three years on mangrove planting at the Pattany Bay. In Proceeding of the 8th National Seminar on Mangrove Ecology, 25-28 August 1993, Surat Thani, Thailand. National Research Council of Thailand, II2: 1-10 (in Thai).

Bamroongrugsa, N. 1997. Mangrove afforestation on newly accreted mudflats of the Pak Phanang Bay, Nakhon Si Thammarat province. In: Proceeding of the 10th National Seminar on Mangrove Ecology, 25-28 August 1997, Hat Yai, Thailand. National Research Council of Thailand, II3: 1-9 (in Thai).

Colbert, K.C. and Larsen, D.R. 2002. Height-diameter equations for thirteen Midwestern Bottomland hardwood species. Northern Journal of Applied Forestry 19: 171-176.

Cole, T.G., Ekel, K.C. and Devoe, N.N. 1999. Structure of mangrove trees and forests in Micronesia. Forest Ecology and Management 117, 95-109.

Devoe, N.N. and Cole, T.G. 1996. Diameter and volume increment in mangrove forest of the Federated States of Micronesia. In: Proceeding of the FORTROP'96: Tropical Forest in the 21st Century, 25-28 November 1996. Kasetsart University, Bangkok, Thailand. pp.57-71.

Dewar, R.C. and McMurtrie, R.E. 1996. Analytical model of stem wood growth in relation to nitrogen supply. Tree Physiology 16: 161–171.

Duarte, C.M., Thampanya, U., Terrados J., Geertz-Hansen, O. and Fortes, M.D. 1999. The Determination of the age and growth of SE Asian mangrove seedlings from internodal counts. Mangroves and Salt Marshes 3: 251-257.

Food and Agricultural Organization (FAO). 1982. Management and utilization of mangroves in Asia and the Pacific. FAO Environment paper no. 3, FAO, Rome, Italy, 160 pp.

FAO. 1993. Forest resources assessment 1990: Tropical countries. FAO forestry paper no. 112, FAO, Rome, Italy. 104 pp.

FAO. 1994. Mangrove forest management guidelines. FAO Forestry paper no. 117, FAO, Rome, Italy, 319p.

Fang, Z. and Bailey, R.L. 1998. Height-diameter models for tropical forest on Hainan Island in Southern China. Forest Ecology and Management 110: 315-327.

Fujimoto, K., Miyagi, T., Kikuchi, T. and Kawana, T. 1996. Mangrove habitat formation and response to Holocene sea-level changes on Kosrae Island, Micronesia. Mangroves and Salt Marshes 1: 47-57.

Ha, T.H., Duarte, C.M., Tri, N.H., Terrados, J. and Borum, J. 2003. Growth and population dynamics during early stages of the mangrove *Kandelia candel* in Halong Bay, North Vietnam. Estuarine, Coastal and Shelf Science 58: 435-444.

Havanond, S., Chukwamdee, J., Anunsiriwat, A. and Meepol, W. 1994. Structure of mangrove Forest at Samut songkram province. Final Report of the Coastal Wetland Conservation Project Wildlife Fund, Thailand under The Royal Patronage of H.M. The Queen, Bangkok, Thailand, 29 pp.

Hoshino, D., Nishimura, N. and Yamamoto, S. 2001. Age, size structure and spatial pattern of major tree species in an old-growth *Chamaecyparis obtuse* forest, central Japan. Forest Ecology and Management 152: 31-43.

Jirawattanapun, S., Patanaponpaiboon, P., Aksornkoae, S. and Pliyavuth, C. 2002. Effects of salinity on distribution of *Sonneratia caseolarlis* and *Soneratia alba*. In: Proceeding of the 12th Thailand National Seminar on Mangrove Ecology, 28-30 August 2002, Nakhon Si Thammarat, Thailand, National Research Council of Thailand, III3: 1-8 (in Thai).

Kaewwongsri, P. and Bamroongrugsa, N. 1997a. Study on Growth of *Avicennia spp.* For pioneer plants on newly formed mudflats of the Pattani Bay, southern Thailand. In Proceeding of the 10th Thailand National Seminar on Mangrove Ecology, 25-28 August 1997, Hat Yai, Thailand. National Research Council of Thailand, II4: 1-8 (in Thai).

Kaewwongsri, P. and Bamroongrugsa, N. 1997b. A study on mangrove vegetation at the Pak Phanang Bay, Nakhon Si Thammarat province. In: Proceeding of the 10th Thailand National Seminar on Mangrove Ecology, 25-28 August 1997, Hat Yai, Thailand. National Research Council of Thailand, II9: 1-17 (in Thai).

Kongsangchai, J., Havanond, S., Jintana, W. and Thanapermpool, P. 1988. The influence of deposited sediment from mines on structure and productivity of mangrove forest in Phang-nag province. Royal Forest Department, Bangkok, Thailand. 50 pp.

Kongsangdow, T., Kooha, B. and Komolsathit, N. 1988. Experiment of pioneer species plantation for mangrove rehabilitation. In: Proceeding of The 6th national seminar on mangrove ecology, 29-31 August 1988, Nakhon Si Thammarat, Thailand. National Research Council of Thailand. pp.212-224 (in Thai).

Lederman, T.C. and Tett, P. 1981. Problem in modeling the photosynthesis-light relationship for phytoplankton. Botanica Marina 14: 125-134.

Lee, S.K., Tan, W.H. and Havanond, S. 1996. Regeneration and colonisation of mangrove on clay-filled reclaimed land in Singapore. Hydrobiologia 319: 23-35.

Loewenstein, E.F., Johnson, P.S. and Garrett, H.E. 2000. Age and diameter structure of a managed uneven-aged oak forest. Canadian Journal of Forest Research 30: 1060-1070.

Maa, C.X., Casellaa, G. and Wua, R. 2002. Functional Mapping of Quantitative Trait Loci Underlying the Character Process: A Theoretical Framework. Genetics 161: 1751-1762.

Macintosh, D.J., Ashton, E.C. and Havanon, S. 2002. Mangrove rehabilitation and intertidal biodiversity; A Study in the Ranong mangrove ecosystem, Thailand. Estuarine, Coastal and Shelf Science 55: 331-345.

McDowell, N., Barnard. H., Bond, B.J., Hinckley, T., Hubbard, R.M., Ishii, H., Köstner, B., Magnani, F., Marshall, J.D., Meinzer, F.C., Phillips, N., Ryan, M.G. and Whitehead, D. 2002. The relationship between tree height and leaf area: sapwood area ratio. Oecologia 132: 12-20.

Menezes, M., Berger, U. and Worbes, M. 2003. Annual growth rings and long-term growth patterns of mangrove trees from Bragaca, North Brazil. Wetlands Ecology and Management 11: 233-242.

Naylor, R.L., Bonine, K.M., Katherine, C.E. and Waguk, E. 2002. Migration, markets, and mangrove resource use on Kosrae, Federated States of Micronesia. Ambio 31: 340–350.

Neter, J., Kutner, M.H., Nachtsheim, C.J. and Wasserman, W. 1996. Applied linear statistical models. 4[th]ed. Time Mirror Higher Education Group, Inc.,1408 pp.

Ott, L. and Longnecker, M. 2001. An introduction to statistical methods and data analysis. Wadsworth Group, 1184 pp.

Panapitukkul, N., Duarte, C.M., Thampanya, U., Kheowvongsri, P., Srichai, N., Geertz-Hansen, O., Terrados, J. and Boromthanarath, S. 1998. Mangrove colonization: Mangrove progression over the growing Pak Phanang (SE Thailand) mud flat. Estuarine, Coastal and Shelf Science 47: 51-61.

Pienaar, F.J. and Turnbull, K.J. 1973. The Chapman – Richards generalization of Von Bertalanffy's growth model for basal area growth and yield in even aged stands. Forest Science 19:2 – 22.

Peper, P.J., McPherson, E.G. and Mori, S.M. 2001. Equation for predicting diameter, height, crown width, and leaf area of San Joaquin valley street trees. Journal of Arboriculture 27 (6): 306-317.

Piriyayotha, S. and Jaicheeng, D. 2001. The appraisement in sustainable and appropriate community mangrove management: A case study of Bang Toey sub-district, Phang Nga Province, Thailand. In *Proceeding of the 11th National Mangrove Seminar, 9-12 July 2000*, Trang, Thailand. National Research Council of Thailand, V8: 1-10.

Plathong, J. and Sitthirach, N. 1998. Traditional and current use of mangrove forest in Southern Thailand. Wetlands International-Thailand Programme/PSU, Publication No.3, 91 pp.

Putz, F.E. and Chan, H.T. 1986. Tree growth, dynamics, and productivity in mature mangrove forest in Malaysia. Forest Ecology and Management 17 (2): 211-230.

Royal Forest Department (RFD). 2001. General field procedures manual. Technical Report No. 4., International Tropical Timber Organization (ITTO) Project "Preparatory studies to install a continuous monitoring system for the sustainable management of Thailand's forest resources", Bangkok, Thailand.

Sandrasegaran, K. 1971. Height-diameter-age multiple regression model for *Rhizophora apiculata* BL. (Syn. *Rhizophora conjugata* Linn.) in Matang mangroves Taiping, West Malaysia. The Malayan Forester 34 (4): 260-275.

Satyanarayana, B., Raman, A.V., Dehairs, F., Kalavati, C. and Chadramohan, P. 2002. Mangrove floristic and zonation patterns of Coringa, Kakinada Bay, East Coast of India. Wetlands Ecology and Management 10:25-39.

Semesi, A.K. 1998. Mangrove management and utilization in Eastern Africa. Ambio 27 (8): 620-626.

Steinke, T.D., Ward, C.J. and Rajh, A. 1995. Forest structure and biomass of mangrove in the Mgeni estuary, South Africa. Hydrobiologia 295: 159-166.

Tabbush, P. and White, J.E.J. 1996. Estimation of tree age in ancient yew woodland at Kingley Vale. Quarterly Journal of Forestry 90: 197-206.

Tam, N.F.Y, Wong, Y.S., Lan, C.Y. and Chen, G.Z. 1995. Community structure and standing crop biomass of a mangrove forest in Futian Nature Reserve, Shenzhen, China. Hydrobiologia 295: 193-201.

Tomlinson, P.B. 1986. The botany of mangroves. Cambridge University Press, 418 pp.

Turner, M., Gong, W.K., Ong, J.E., Bujang, J.S. and Kohyama, T. 1995. The architecture and allometry of mangrove saplings. Hydrobiologia 9: 205-212.

Verheyden, A., Kairo, J.G., Beeckman, H. and Koedam, N. 2004. Growth ring, growth ring formation and age determination in the mangrove *Rhizophora mucronata*. Annals of Botany 94: 59-66.

Wechakit, D. 1987. Growth and yield of *Rhizophora apiculata* planted in private forest, Samut Songkram province, Thailand. M.Sc. Thesis. Kasetsart University, 71pp.

Zeide, B. and Vanderschaaf, C. 2002. The effect of density on the height-diameter relationship. In: Outcalt, K.W. (ed), Proceeding of the eleventh biennial southern silvicultural research conference. Gen. Tech. Rep. RSR-48. Asheville, NC: U.S. Department of Agriculture, Forest Service, Southern Research Station, 622 pp.

Annex

Annex 4.1. Regression comparison using dummy variables for height-age and DBH-age. Full models were compared to reduced models both in multiple and pair-wise comparisons. Presented are F computed (F), degree of freedoms of numerator (df_1) and denominator (df_2), and significant level (p). Age is independent variable.

comparison	dependent variable	slope				intercept			
		F	df_1	df_2	p	F	df_1	df_2	p
within *R. apiculata*	height	1.35	2	39	0.272	7.94	2	40	<0.001
	DBH	4.41	2	32	0.020	13.85	2	34	<0.001
among *Avicennia*	height	5.19	2	30	0.012	9.26	2	32	<0.001
	DBH	5.62	2	20	0.012	10.42	2	22	<0.001
between *Rhizophora*	height	0.10	1	34	0.759	13.89	1	35	<0.001
	DBH	11.43	1	28	0.002	104.17	1	29	<0.001
within *R. apiculata*									
Pattani vs Ranong	height	1.80	1	25	0.192	2.13	1	26	0.157
	DBH	0.13	1	21	0.725	37.09	1	22	<0.001
Ranong vs	height	0.002	1	26	0.964	7.61	1	27	0.010
Samut Songkram	DBH	4.91	1	21	0.038	16.44	1	22	0.001
Samut Songkram vs	height	0.63	1	25	0.435	9.62	1	26	0.005
Pattani	DBH	6.27	1	22	0.020	0.90	1	23	0.352
among *Avicennia*									
A. alba vs *A. marina*	height	0.15	1	21	0.704	11.94	1	22	0.002
	DBH	14.82	1	14	0.002	7.10	1	15	0.018
A. alba vs *A. officinalis*	height	7.18	1	19	0.015	9.83	1	20	0.005
	DBH	1.26	1	13	0.283	22.66	1	14	<0.001
A. marina vs	height	13.83	1	20	0.001	0.01	1	21	0.919
A. officinalis	DBH	4.80	1	13	0.047	3.54	1	14	0.081

Chapter 5

Long-term mangrove progression and coastal recession along the coasts of Southern Thailand

Udomluck Thampanya, Jan E. Vermaat and Sin Sinsakul

Abstract

Approximately sixty percent of the southern Thai coastline used to be occupied by mangroves according to the first mangrove forest assessment in 1961. During the past three decades, these mangrove areas have been reduced to about fifty percent with less than ten percent left on the east coast. Coastal erosion and accretion occur irregularly along the coast but an intensification of erosion has been noticed during the past decade.

This study assessed the relationship between mangrove presence and changes in coastal area. Mangrove colonization rates were assessed using both *in situ* transects and remote sensing time series. Both methods led to comparable estimates ranging between 5 and 40 m y^{-1}. Quantitative data on changes of coastal segments along the Gulf of Thailand and Andaman Sea coasts as well as available possible factors responsible for these changes were compiled. Overall, net erosion prevailed (1.3 ± 0.4 m y^{-1}). The Gulf coastline in the East of the country was found to be most dynamic: change occurred along more coastal segments than in the West (43% vs 16%). Rates of erosion and accretion were also higher, 3.6 versus 2.9 m y^{-1} and 2.6 versus 1.5 m y^{-1}, respectively. Total area losses accounted for 0.91 km^2 y^{-1} for the Gulf coast and 0.25 km^2 y^{-1} for the West. Coasts with and without mangroves behaved differently: in the presence of mangroves less erosion was observed whilst expansion occurred at particular coastal types with mangrove existence, i.e. river mouths and sheltered bays. Possible underlying causes were examined using multivariate analysis. Eroded areas were found to increase with increased area of shrimp farms, increased fetch to the prevailing monsoon, and when dams reduced riverine inputs. Notably, however, in areas where erosion prevailed, the presence of mangroves reduced these erosion rates. Mangrove loss was found to be higher in the presence of shrimp farms and in areas where mangrove forests used to be extensive in the past.

Submitted to Estuarine, Coastal and Shelf Science and has been accepted for publication in January 2006.

Introduction

Historically, large tracts of the coastal zone of SE Asia have been occupied by mangroves (Rao 1986; Aksornkoae 1993). During the past decades, these mangroves have been cleared over vast areas to accommodate increasingly intensive forms of land-use for human benefit such as settlement, transport infrastructure, agriculture and aquaculture, especially shrimp farming. Traditionally, mangrove forests provide the coastal human population with a variety of goods and services on which poorer strata of society depend strongly (Aksornkoae 1993; Ruitenbeek 1994; Plathong & Sitthirach 1998; Gilbert & Janssen 1998; Semesi 1998; Janssen & Padilla 1999).

The notion of the importance of mangrove forests has urged widespread reforestation schemes to cope with this decline (Havanond 1995; FAO 1994; Field 1999). The success of reforestation has been variable, however, amongst others due to the neglect of the ecology of sites and species (Khemnark 1995; Havanond 1995; Elster 2000; Thampanya et al. 2002a, Chapter 2; Thampanya et al. 2002b, Chapter 3). Mangroves have their widest extent in lowland deltas (Woodroffe 1992; Robertson & Alongi 1992) i.e. where sediment delivery allows a net progression of a soft-bottom coastline. Deltaic coasts are affected by changes both on the land and in the sea. Anthropogenic activities in upland catchments such as deforestation, cultivation, dam constructions as well as coastal activities such as construction of ports, sand barriers, break-waters and jetties, all may have adverse impacts on the sediment delivery and thus on the availability of mangrove habitats (FAO 1994; Saito 2001; Hogarth 2001). An important question is whether mangroves simply follow the geologically changing coastlines or also protects the coastlines from erosion, hence accelerate the entrapment of suspended particulate matter from land and sea (Thom 1982; Field 1995; Blasco et al. 1996; Furukawa & Wolanski 1996). The latter would provide a potentially powerful feedback enabling the continued existence of mangrove stands at reduced terrestrial sediment delivery and, possibly, foreseen sea level rise (Ellison 1993; Blasco et al. 1996; Hogarth 2001). However, it is difficult to establish experimentally, because of the vast spatial scale of the system invaded. We therefore attempted a multivariate approach, using provinces and coastal segments in Southern Thailand as replicates.

In this paper we provide a regional overview of coastal progradation and erosion along the coasts of Southern Thailand and specifically focus on long-term development in a number of areas with comparatively large mangrove stands. We combine detailed longer time coastal development assessment for these sites using remote sensing in combination with *in situ* studies and compare the results with overall patterns synthesized from coastal surveys for all coastal provinces of Southern Thailand (from Sinsakul et al. 1999; 2002). Our aims were:

(1) to assess whether natural coastal development of mangrove-dominated coastlines is different from that of others; and

(2) to identify factors responsible for the expansion or recession of mangrove-dominated coastlines.

In addition, we address a methodology issue and verify whether remote sensing and *in situ* tree-age size distributions (Panapitukkul et al. 1998) provide comparable estimates of coastline progradation.

Materials and methods

Study area

Southern Thailand is approximately situated between 6° to 11° Northern latitude and 98° to 103° Western longitude. Its coastlines face two different seas: the eastern is exposed to the Gulf of Thailand and the western to the Andaman Sea. The Gulf of Thailand, an inlet of the South China Sea, has a coastline along the southern region stretching for approximately 930 km. Many rivers discharge water and sediment into the gulf. The Andaman Sea, which connected to the Indian Ocean, has a 937 km long coastline.

The geomorphology of the gulf coast is characterized by a long and wide mainland beach of sand and dunes, with lagoons, bays and spits. Pocket beaches, extensive and well-preserved tidal flats, cliff coasts and numerous islands dominate the Andaman coast. Extensive mangroves along the Andaman coast accounted for approximately eighty percent of total mangrove area (1,675 km^2) of the country in 1998 (Royal Forest Department 2004). In the gulf coast, the remaining mangroves are present at sheltered coastlines (Figure 4.1). Tidal range on the east coast varies slightly between 0.3-1.1 m with two types of tide: mixed but predominantly diurnal in the upper part and mixed but predominantly semidiurnal in the lower part. Along the western coast, the tide is mixed semidiurnal with a relatively high tidal range between 1.1 –3.6 m (Siripong 1985).

The area has a tropical climate with two monsoonal winds: the northeast (NE) during mid October to March and the southwest (SW) during May to September. The NE wind has a longer fetch and mainly generates waves along the Gulf coast. Highest waves along the Andaman coast are generated by the SW monsoonal wind. Peaks in wind and wave intensity caused by the passing cyclones frequently accompany the retreat of the monsoon during October to November (Vongvisessomjai et al.1996). The annual rainfall of the Southern region is higher than in other parts of Thailand and highest precipitation occurs on the Andaman coast (2,100 to 4,000 mm y^{-1}) whilst it ranges between 1,600-2,400 mm y^{-1} on the Gulf coast.

There are six coastal provinces along the east coast and another six along the west coast. In this study, we selected four provinces with extensive mangrove areas as the sites for our detailed surveys of mangrove progression: Ranong and Satun on the Andaman coast and Nakhon Si Thammarat and Pattani on the Gulf coast (Figure 5.1). Different mangrove taxa prevail in these four sites (Table 5.1).

Table 5.1. Abundance of main mangrove species present along the mangrove forest edges in the selected sites (based on field observations and Plathong & Sitthirach 1998).

Species	Sites			
	Nakhon Si Thammarat	Pattani	Ranong	Satun
Avicennia alba	a	a	f	f
Avicennia officinalis	f	o	o	o
Avicennia marina	f	a	n	n
Rhizophora apiculata	a	a	a	a
Rhizophora mucronata	f	f	o	f
Sonneratia alba	o	o	f	f
Sonneratia caseolaris	a	o	o	o

a = abundant; f = frequent; o = occasionally; n = not present

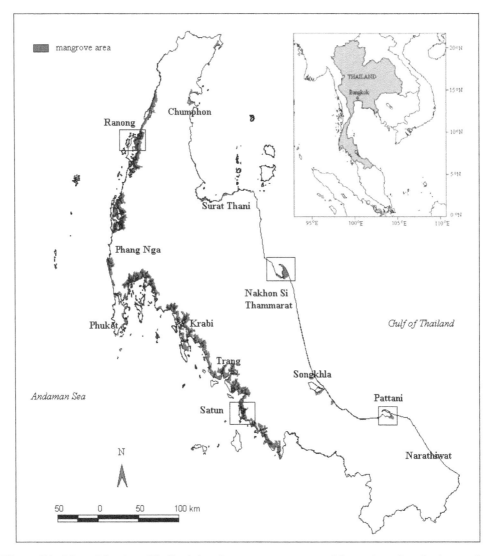

Figure 5.1. Map of Southern Thailand showing mangrove areas and four selected study sites, and geographic names use in the text.

Remote sensing method

To distinguish changes in mangrove forest expansion and cover longer-time development, we use a 30 years time period for assessment. Therefore, black and white aerial photographs and Landsat-TM satellite data of the four study sites were used. The aerial photographs of 6 and 7 September 1966 (resolution of 1:50,000, analog format, dimension: 23 cm x 23 cm) were available for Satun and Nakhon Si Thammarat, respectively, while black and white aerial photographs of 23 March 1967 were obtained for Ranong and 1 April 1967 for Pattani. Available cloud free Landsat-TM imagery (pixel size of 25 m, band 1-7, digital format) of Satun and Pak Phanang areas were acquired on 7 March 1997 and 1 October 1997 while those of Pattani and Ranong were acquired on 20 April 1998 and 24 August 1998, respectively. In addition, a topographic map of each area, with resolution of 1:50,000 in analog format produced by the Royal Thai Survey Department in 1973, was used as reference material.

All analog data were transformed into digital data using scanners. Data preparation and image processing were carried out using the ERDAS Imagine software version 8.5. The aerial photo image of each study area was geo-referenced with the topographic map considering the UTM-Everest co-ordinates system and thus registered with the satellite image. In addition, it is necessary for aerial photographs to be joined together under mosaic operation to cover the area of study while satellite images were subset. Since this study emphasizes mangrove forest edges, areas in the vicinity of the shoreline have been examined.

The visible bands of satellite data (RGB 3,2,1: true colour composite) were used for a first reconnaissance. Due to a relatively low image quality, image enhancement technique was employed to the aerial photograph of Ranong. Thereafter, both aerial photographs and satellite images were classified using unsupervised classification with 10 classes. Subsequently, the subclasses were merged and recoded to three classes: water, mangroves and mud. Finally, the classified aerial photo image and satellite image were added (Panapitukkul et al. 1998) by operation utility.

Mangrove expansion is interpreted from the resulting pictures as differences in mangrove forest edge and area between the two successive dates. Between 40-80 perpendicular lines were drawn between the two edges and progression was measured from these lines as distances between 1966 to 1997 for Satun and Nakhon Si Thammarat and 1967 to 1998 for Pattani and Ranong.

Ground Truthing

The reconnaissance surveys of the study areas have been made within the Coastal Ecosystems Response to Deforestation-derived Siltation in Southeast Asia (CERDS) Project during 1996-1997 (Panapitukkul et al. 1998; Kamp-Nielsen et al. 2002). The results of image classification from the remote sensing (RS) methodology were verified based on these surveys as well as from later *in situ* fieldwork by the first author. The *in situ* studies were carried out at all sites (Pattani in September 1999, Ranong and Satun in August 2000 and, Nakhon Si Thammarat in September 2000) in order to quantify mangrove expansion rates and validate the progression rates obtained from the remote sensing technique. Temporal line transects were set up in the areas where mangrove progression was apparent according to the remote sensing results. A Global Positioning System (GPS) was used to locate the transects.

Transect lines started from the edge of the present closed forest characterized with mature trees perpendicular to the waterway or to the furthermost individual. Then 10x10 m plots were set up evenly along each transect to examine age of mangrove trees. Number of line transects and number of plots varied among the four sites depending on their mangrove characteristics and accessibility. In Pak Phanang Bay, Nakhon Si Thammarat, three line transects were set up in bay features and another four in cape features with 8-10 plots on each transect. For the other study sites, mangrove progression areas are small, therefore, five transects with 5-10 plots each were located for Ranong and Pattani Bay and eight transects with the same number of plots were located for Satun area.

In each plot, we measured height and circumference at breast height (1.3 m from substrate) of the three largest specimens. Subsequently, data on girth at breast height were converted to diameter at breast height (DBH) using circumference = π x DBH, where π=3.14. For young trees with height less than 2 m, we also counted the number of internodes present along the main stem as in Panapitukkul et al. (1998). Age of the oldest tree was estimated using fitted allometric height-age or DBH-age curves (Thampanya et al. in prep, Chapter 4). For young trees, age was calculated from the ratio of total number of internodes and the average annual number of internodes produced for each species i.e. 16.0±0.8 internodes for *Avicennia*, 25.0±1.2 for *Sonneratia caseolaris* and 8.3±0.4 for *Rhizophora apiculata* (Duarte et al. 1999; see also Chapter 2).

The maximum age of the oldest individual along the transect represents an estimated time since that tree colonized the plot (Panapitukkul et al. 1998) and also of the time that the mudflat emerged sufficiently high above the low water line to be colonized. Hence, the progression rate of mangrove colonization was estimated from the fitted regression equation derived from the increase distance from the present forest edge in time, as follows:

$$\text{Distance from forest edge (m)} = A + B \times \text{Maximum age (days)}$$

Where:

A = intercept of the regression equation
B = slope or progression rate (m day^{-1})

Compilation of datasets

To obtain an overview of coastal change patterns in Southern Thailand and assess underlying factors, datasets on natural coastal development of this region and possible causal factors were compiled. The results of coastal surveys by the Geological Survey Division, Department of Mineral Resources, Thailand (Sinsakul et al. 1999; 2002), were taken into consideration. These surveys aimed to assess the current status of the coastal geo-environment, coastal problems and their causes. The Andaman Sea coast was surveyed in the year 1997. For the Gulf of Thailand, the survey was carried out during 1998-2000.

We synthesized relevant quantitative data from these surveys into a dataset of coastal development for all coastal provinces of Southern Thailand. This set contains data on coastal change during 1967-1998 in terms of erosion and accretion for coastal segments of 0.5-30 km long (n=142), magnitude (rate in m y^{-1}), cause of change, coastal type and, presence and condition of mangroves in the areas. We used six categories of coastal type: rocky, rocky sand, sandy, sandy mud, muddy, and river/canal mouth. Similarly, four categories were assigned to the presence and condition of mangroves: dense or intact, scattered or degraded, former mangrove area and not present.

Comprehensive long-term baseline oceanographic data such as wave, wind and current of this region are limited. This study, therefore, attempted to compile quantitative data of other direct or indirect possible causal factors for coastal change from various sources. Some important possible factors were also calculated to accomplish the dataset. This set consists of southern Thai data on a province basis including precipitation, hydrological, land use and important coastal data such as coastal length and fetch.

Data analysis

The data on mangrove expansion and coastal change were subsequently analyzed using standard statistical software (SPSS and Microsoft Excel). Analysis of variance was applied to examine the difference of mangrove progression rate resulting from the RS method and ground truthing as well as to examine the effect of coastal factors such as locality, coastal type and mangrove presence, on rate of coastal change.

Possible factors responsible for coastal change were assessed using stepwise multiple regressions. We examined these for net coastal change, coastal erosion and accretion subsequently. In addition, mangrove area loss was also examined for its causal factors. Since no mangroves have been reported for Narathiwat province, this province was not selected in the data analysis concerning mangrove aspects.

Results

Mangrove progression in selected sites

Remote sensing imagery suggested that mangrove progression differed strongly between eastern and western sites (Figure 5.2). In the east, mangrove progress in the selected river mouths has been in the order of 140 to 1200 m over the period of about 31 years, or approximately 5 to 38 m y^{-1}. In the west, however, equally large mangrove forests increased only little in the seaward direction: 10 to 90 m over 31 years or less than 3 m y^{-1}. Mean annual progression rate was significantly higher in Pak Phanang Bay, Nakhon Si Thammarat than in the other three sites (Figure 5.2 and 5.3, Table 5.2).

Progress measured from remote sensing imagery and ground truth transects were compared (Table 5.2). No significant difference was found in overall progression rate estimates by the two methods ($p=0.125$), but the method and site interaction was significant (Table 5.2). This can probably be explained by the fact that the difference between the two methods was relatively large at one site, Ranong, where the *in situ* transects still suggest some progress (3 ± 1 m y^{-1}) but the remote sensing method showed little progress (0.3 ± 0.03 m y^{-1}, Figure 5.3). In the other sites both methods yielded rather comparable estimates of mangrove progression.

West coast   East coast

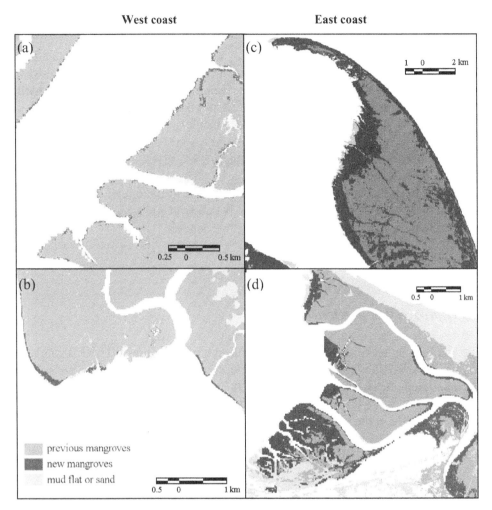

Figure 5.2. Total mangrove progression observed from remote sensing imagery over 31 years at four selected sites on the coast of Southern Thailand. (a) Ranong, (b) Satun, (c) Nakhon Si Thammarat and (d) Pattani. Pictures have different scales.

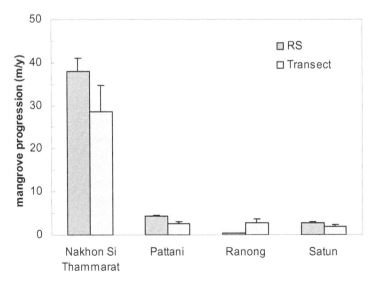

Figure 5.3. Mangrove progression rates obtaining from the two methods: RS (remote sensing) and transect (ground truthing), and the Tukey comparison of the four study sites..

Table 5.2. Univariate Analysis examining the effects of sites and method on mangrove progression rate. Presented are the degree of freedom (*df*), percentage of variance explained (factor SS/total SS x100) and the level of significant (*p*).

factors	*df*	% variance	*p*
Method	1	<2	0.125
Site	3	57	0.000
Method x Site	3	2	0.038
Residual	227	39	
Total	235	100	

Erosion and accretion along southern Thai coastlines

It was found that all coastal provinces of Southern Thailand, in the east as well as the west, have been experiencing coastline change. Retreats of marginal shoreline occurred irregularly. Erosion was more prevalent on the Gulf coast than the Andaman coast. For instance, large lengths of the eastern coastline were found to be subject to erosion, particularly in Nakhon Si Thammarat and Narathiwat, whilst on the western coast this is less pronounced. On the other hand, tidal flats and sand bars in the East have also been prograding continuously with substantial rates in sheltered areas such as Pak Phanang Bay and Pattani Bay. In general, however, coastal areas of Southern Thailand are diminishing. Area losses amount from approximately 0.01 to 0.32 $km^2 y^{-1}$ per province (Table 5.3).

Table 5.3. Coastal change in the coastal provinces of Southern Thailand, area of mangroves, number of coastal segments[a] with changed area (N), length of changed coastline (L, km), mean rate (SE) of change (R, m y^{-1}) and estimated net change over the period of 1967-1998.

province	length of coastline (km)	mangrove area in 2000 (km^2)	erosion N	erosion L	erosion R	accretion N	accretion L	accretion R	estimated net change ($km^2.y^{-1}$)
East coast									
Chumphon	185	79	11	16	1.6 (0.1)	8	10	1.6 (0.2)	-0.01
Surat Thani	135	35	8	24	3.5 (1.8)	9	11	1.6 (0.2)	-0.14
Nakhon Si Thammarat	190	99	9	112	4.0 (0.9)	4	21	8.9 (7.0)	-0.32
Songkhla	150	47	11	41	2.3 (0.4)	4	32	2.5 (0.3)	-0.02
Pattani	170	35	8	24	6.7 (1.7)	3	6	2.8 (0.7)	-0.12
Narathiwat	50	0	6	41	5.0 (1.6)	1	2	4.0 (0.0)	-0.29
Total	880	296	53	258	3.6 (0.5)[b]	29	82	2.9 (1.0)[b]	-0.13[b]
West coast									
Ranong	135	253	7	25	2.6 (0.6)	1	4	2.0 (0.0)	-0.06
Phang Nga	216	454	9	28	1.8 (0.2)	5	10	1.5 (0.3)	-0.04
Phuket	185	22	2	5	5.8 (4.3)	0	-	-	-0.03
Krabi	160	349	9	17	2.7 (0.6)	5	6	1.6 (0.4)	-0.03
Trang	119	335	7	23	2.2 (0.5)	5	11	1.0 (0.0)	-0.05
Satun	168	353	9	15	3.0 (0.8)	1	4	2.5 (0.0)	-0.03
Total	983	1,766	43	113	2.6 (0.3)[b]	17	35	1.5 (0.2)[b]	-0.04[b]

[a] segment is a portion of coastline in a province which was observed to be eroded or accreted; [b] average rate

Different provinces along Southern Thai coastlines displayed great contrasts in terms of percent coastline change (Figure 5.4). Overall, eroded distances for the east and the west coasts accounted for 29% and 11% of their total length, respectively. Narathiwat, the southernmost province, exhibited the highest proportion of eroding coastline (82% of its 50 km) followed by 59% for Nakhon Si Thammarat (Figure 5.4), while a maximum of only 19% was found on the west coast (Ranong and Trang). Percent accretion had a similar pattern, ranging from 2% to 9% for the west and 3% to 21% for the east coast.

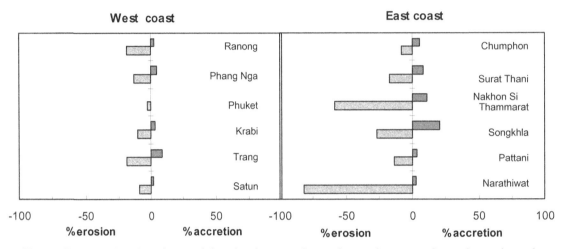

Figure 5.4. Percent of total coastal length where erosion and accretion occur for each province along Southern peninsular Thailand.

Although the west and east coasts they appear to show different patterns in percentage of coastal change among provinces (Figure 5.4), in an overall analysis of variance considering rate of change (both erosion and progradation) among segments, we found no significant difference between the two coasts as well as among the six different types of coastline distinguished (Table 5.4), probably due to the large variability observed. In contrast, the degree of mangrove presence did matter since it was highly significant explaining about 11% of the variance ($p<0.001$, Table 5.4). Additionally, the interaction between degree of mangrove presence with coast and with coastal type showed significant differences (Table 5.4). This suggests that coastlines with mangroves behave differently on the Andaman and on the Gulf coasts, as well as on different coastal types.

Table 5.4. Analysis of variance examining the effects of coastal side (eastern, western), coastal type and presence of mangroves on rate of coastal change (both negative erosion and positive accretion) in 142 coastal segments. Presented are the degree of freedom (*df*), SS, percentage of variance explained (factor SS/total SS X 100), and level of significance (*p*).

factors	df	SS	% variance	p
Coast (west vs. east)	1	2	<1	0.709
Coastal types[a]	5	49	2	0.625
Degree of mangrove presence	3	327	11	0.000
Coast X Coastal type	3	98	3	0.078
Mangrove presence X Coast	3	191	7	0.005
Mangrove presence X Coastal type	5	270	9	0.003
Residual	121	1690	59	
Total	142	2868		

[a]coastal types distinguished = see figure 5.5.

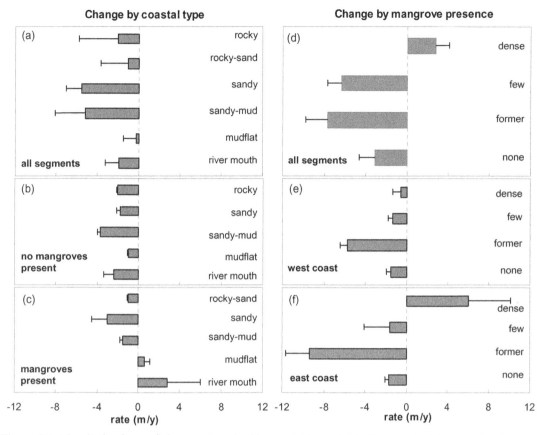

Figure 5.5. Magnitude of coastal change categorized by coastal type of: (a) all segments, (b) segments with no mangrove presence, (c) segments with mangrove presence; and degree of mangrove presence of: (d) all segments, (e) segments of the west coast, and (f) segments of the east coast.

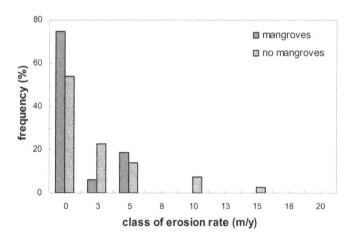

Figure 5.6. Class of erosion rate and frequency of coastal segments with and without mangroves. Mean erosion rates are 2.3 ± 0.4 m y^{-1} for segments with mangroves and 3.3 ± 0.4 m y^{-1} for segments without mangroves. The t-test statistics shows significant different erosion rate between these two coastal types (t-test, p=0.48).

Remarkably, in the overall ANOVA we found no significant difference in rate of coastal change among the six coastal types despite the apparent pattern in the means (Figure 5.5 a). We must conclude that variability within these types is substantial along Southern Thai coastlines. Probably, sandy and sandy-mud coasts are more fragile than the others whilst mudflats experienced less erosion (less than one m y^{-1}). This is also in accordance with the presence of mangroves (Figure 5.5d), which usually dominate mudflat shorelines and river mouths. Here too significant interactions between mangrove presence and coast (east-west) as well as mangrove presence and coast type (Table 5.4) suggest the importance of mangrove presence: only where mangroves are present, accretion is positive (Figure 5.5c) and accretion occurs only where dense mangroves occur along the east coast (Figure 5.5f). In addition, the erosion rate was found to be higher in coastal segments without mangroves than those with mangroves (Figure 5.6).

Possible factors responsible for coastal changes

Generally, there are many inter-related causes of coastal change, which can be long-term and short-term, natural and man-induced. In this study, we collected relevant data on possible factors responsible for coastal changes from several sources as summarized in Table 5.5 where we used province as unit of comparison. Available long-term data were averaged to get representative values. Data on rainfall were available for the period of 1971-2000 and sediment yield for 1979-2000. The examined coastal change data covered the period between 1967 and 1998.

Table 5.5. Ranges across provinces in quantified possible causal factors of southern Thai coastal change used in multiple regression analysis.

variable no./ name	median	min	max	sources
1. coastal length (km)	164	50	216	Sinsakul et al. 1999; 2002
2. catchment area (km^2)	2,838	159	8,969	Royal Irrigation Department 2004
3. rainfall (mm y^{-1})	2,197	1,816	4,021	Meteorological Department 2004
4. sediment yield (ton/km^2)	160	85	339	Royal Irrigation Department 2004
5. presence of dams in major rivers' draining a province (yes/no)	-	-	-	Royal Irrigation Department 2004
6. coastal water suspended solids in 2003 (mg l^{-1})	17	7	46	Pollution Control Department 2003
7. river plume (km^2)	102	24	477	estimated from 2000 and 2001 Landsat Satellite images
8. mangrove area in 1961 (km^2)	281	13	612	Royal Forest Department 2004
9. mangrove area in 1996 (km^2)	58	6	304	Royal Forest Department 2004
10. mangrove area loss (km^2)	131	7	528	Royal Forest Department 2004
11. fetch (x100 km)	6	4	19	calculated from world atlas map
12. shrimp farm area in 1996 (km^2)	11	2	105	Charuppat & Charuppat 1997
13. eroded area (x1000 m^2)	64	26	560	Sinsakul et al. 1999; 2002
14. accreted area (x1000 m^2)	12	0.0	240	Sinsakul et al. 1999; 2002
15. eroded distance (x1000 m^2)	24	5	112	Sinsakul et al. 1999; 2002
16. accreted distance (x1000 m^2)	8	0	32	Sinsakul et al. 1999; 2002
17. net area loss (ha) for the period of 1967-1998	36	0.3	290	calculated from Sinsakul et al. 1999; 2002

In stepwise multiple regressions assessing possible causes of coastal change, we have tested a number of models. Two initial models were selected for examining causal factors of net coastal area loss: model (a) examined variable 1 to 14 and model (b) examined variable 1 to 16 (Table 5.6). Stepwise regression then iteratively selected the best (in terms of explain variance) combination of independent variables. Apart from the two obvious factors that add up to net change, i.e. erosion and accretion, the models suggest that fetch, catchment area, coastal length and the presence of dams affected net coastal change significantly. For example, an increase of 100 km fetch enhanced net area loss in the order of 4,580 m^2 (or approximately 0.05 ha), the presence of dams across inflow rivers caused considerable net area loss of about 91,510 m^2 (or 0.92 ha) and loss increased by 10 m^2 for 1 km^2 increase in catchment area (Figure 5.7).

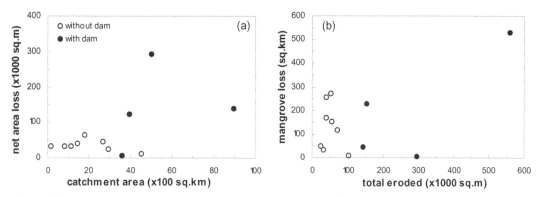

Figure 5.7. (a) net area loss versus size of catchment with and without the presence of dams; (b) total eroded area versus mangrove area loss with and without the presence of dams.

Model (c), (d) and (e) determined correlations of eroded area. In model (c), variable 1 to 11 were included and model (d) examined the same set of variables with the addition of shrimp farms. These two models show that mangrove loss, mangrove area in 1996, fetch and shrimp farm area were factors correlating with coastal erosion. One km^2 of mangrove area loss contributed to 830 m^2 of eroded area and 1 km^2 increment in shrimp farm area increased the eroded area in the order of 2,170 m^2 (or 0.02 ha). On the contrary, 1 km^2 of mangrove area in 1996 reduced the eroded area by 430 m^2. In model (e), the same set of variables was examined but included data of Narathiwat. Here it was found that on top of the mangrove area loss and fetch factors, the presence of dams was significant (a slope of 114x1000 m^2, Table 5.6).

Similarly, shrimp farm area and fetch were significant positive factors of coastal accretion determining in model (f) that examined variable 1 to 12. Here, exposed provinces with more shrimp farming were probably subject to both more erosion (model (d)), and accretion (model (f)) in accordance with figure 5.4. This model also indicates that accreted area may be reduced by 10 m^2 with 1 km^2 increment of catchment's size, or larger catchments witnessed less accretion. In addition, when the eroded area variable was included in model (g), it was found that the presence of dams and amount of rainfall had negative correlation to accreted area (Table 5.6). Model (h) examined the same set of variables with the inclusion of Narathiwat data. This model reveals that apart from the eroded area and the presence of dams, coastal length was another causal factor. An increase of 1 km coastal length led to 470 m^2 of accreted area (Table 5.6).

Table 5.6. Stepwise multiple regressions relating coastal changes to possible causal factors across Southern Thai provinces. Different run (a-j) involved the correlate combination of net loss, eroded area, accreted area and mangrove loss with different sets of independent variables. All models were highly significant. Presented are dependent variables, selected variables, slope and significant level of each selected variable, model intercept, coefficient of determination (r^2) and excluded variables.

model no./ dependent variable	selected variables variable	slope (SE)	p	Intercept (SE)	overall r^2	excluded variable
Determination of net coastal area loss						
(a) net area loss	accreted area (14)	-1.96 (0.13)	0.00	-7.75 (9.36)	0.98	1,3,4,5,6,7,
	eroded area (13)	0.67 (0.06)	0.00			8,9,10,12
	fetch (11)	4.58 (1.01)	0.00			
	catchment area (2)	0.01 (0.002)	0.03			
(b) net area loss	coastal length (1)	-0.58 (0.22)	0.03	103.58 (45.31)	0.95	2,3,4,6,7,8,
	dam (5)	91.51 (17.41)	0.00			9,10,11,12,
	accreted area (14)	-1.58 (0.43)	0.01			13,16
	eroded distance (15)	2.68 (1.05)	0.04			
Determination of coastal erosion						
(c) eroded area	mangrove loss (10)	0.83 (0.09)	0.01	-39.08 (32.37)	0.95	1,2,3,4,5,6,
	fetch (11)	9.06 (2.51)	0.01			7,8,
	mangrove 96 (9)	-0.43 (0.13)	0.01			
(d) eroded area	shrimp farm (12)	2.17 (0.65)	0.01	-90.67 (25.12)	0.95	1,2,3,4,5,6,
	fetch (11)	11.56 (2.21)	0.00			7,8,9
	mangrove loss (10)	0.33 (0.13)	0.04			
(e) eroded area	dam (5)	113.66 (48.91)	0.04	-89.72 (38.89)	0.88	1,2,3,4,6,7,
	mangrove loss (10)	0.48 (0.13)	0.01			8,9,12
	fetch (11)	12.07 (3.74)	0.01			
Determination of coastal accretion						
(f) accreted area	shrimp farm (12)	1.84 (0.15)	0.00	-21.64 (7.78)	0.98	1,3,4,5,6,
	fetch (11)	5.48 (0.67)	0.00			7,8,9,10
	catchment area (2)	-0.01 (0.002)	0.00			
(g) accreted area	eroded area (13)	0.37 (0.02)	0.00	-15.02 (5.43)	0.99	1,2,4,6,7,
	dam (5)	-64.83 (3.21)	0.00			8,9,10
	fetch (11)	3.43 (0.31)	0.03			
	shrimp farm (12)	0.57 (0.09)	0.00			
	rainfall (3)	-0.01 (0.002)	0.01			
(h) accreted area	eroded area (13)	0.51 (0.07)	0.00	-86.68 (29.70)	0.91	2,3,4,6,7,8,
	coastal length (1)	0.47 (0.18)	0.03			9,10,11,12
	dam (5)	-54.79 (22.71)	0.04			
Determination of mangrove area loss						
(i) mangrove loss	mangrove 61 (8)	0.43 (0.04)	0.00	-28.50 (12.58)	0.98	1,2,3,4,5,
	shrimp farm (12)	1.44 (0.47)	0.02			6,7,11
	eroded area (13)	0.25 (0.10)	0.04			
(j) mangrove loss	mangrove 61 (8)	0.56 (0.08)	0.00	-38.73 (30.15)	0.89	1,2,3,4,6,
	dam (5)	131.71 (37.31)	0.01			7,11

Examined variable codes: 1 = coastal length; 2 = catchment area; 3 = rainfall; 4 = sediment yield; 5 = presence of dams; 6 = suspended solid; 7 = river plume; 8 = mangrove area 1961; 9 = mangrove area 1996; 10 = mangrove area loss; 11 = fetch; 12 = shrimp farm area; 13 = eroded area; 14 = accreted area; 15 = eroded distance; 16 = accreted distance. Narathiwat province was not included in runs (c), (d), (f) and (g). In runs (i) and (j), presence of dams and shrimp farms could not be included together as that led to over parameterization.

Factors responsible for mangrove area loss were assessed in model (i) and (j). In model (i), variable 1 to 8 and 11 to 13 were examined. The model suggests that besides the mangrove area in 1961, shrimp farm area and eroded led to mangrove area loss during 1961 to 1996. The shrimp farms was the most severe factor; increases in 1 km^2 of shrimp farm area caused 1.44 km^2 of mangrove area loss, whilst 1,000 m^2 of eroded area related to mangrove area loss of about 0.25 km^2 (Table 5.6). The presence of dams was explicit as one of a possible causal factor of mangrove area loss in model (j) when shrimp farm area and eroded area were excluded from the examined variables. This model suggests that the presence of dams intensified mangrove area loss (Table 5.6, see also Figure 5.7).

Discussion

Generally, mangrove progress estimated in the remotely sensed data and in ground truth transects was quite similar. Only one site (Ranong), where the steep geomorphology prevented the formation of extensive mudflats, showed a significant difference between these two methods. Mangrove expansion rates were in the order of 2-4 m y^{-1}, with the exception of the more rapidly expanding forest in an infilling sheltered bay of Pak Phanang (25-40 m y^{-1} or approximately 0.35 km^2 y^{-1}). This progress rate is, however, considerable less than that reported for the Segara Anakan area in Indonesia (2 km^2 y^{-1}, Purba 1991). The significant difference in progress rate between the two methods at Ranong is probably caused by progress being less than the resolution of the Landsat satellite imagery (pixel size of 25m x 25m) since the majority of measured distances were within 1-2 pixels. Here, the remote sensing method encountered an obvious technical limitation compared to *in situ* transects. A similar limitation was found by Maged et al. (1998).

On the larger spatial scale of coastal segments (0.5-30 km) grouped within 12 coastal provinces, we found overall net coastal erosion. This erosion was significantly higher along the eastern coast than in the west. Presence of mangroves was associated with the remaining areas of positive accretion. Notably, river mouths with mangroves and dense mangrove sites of the east coast were these areas of positive coastal accretion. Elsewhere, irrespective of coastal type, erosion predominated. Erosion rate varied locally between 1.6 to 6.7 m y^{-1} whilst accretion ranged from 1.0 to 8.9 m y^{-1}. Along the east coast, total eroded area amounted to 0.91 km^2 y^{-1} and an average per province was 0.13 m y^{-1} whilst corresponding estimates in the west were 0.25 and 0.04 m y^{-1}, respectively. The overall erosion rate on the east coast is 3.6 m y^{-1}, which is comparable to that of Kuala Terenganu on eastern peninsular Malaysia (0.2-4.0 m y^{-1}, Maged & Mansor 1998). The net area loss found here was higher than that reported for the Bay of Bengal (0.65 km^2 y^{-1}, Ghosh et al. 2001) but slightly less than that of the Camau Peninsula in Vietnam (1.1 km^2 y^{-1}, Saito 2001). Hence, comparable net erosion rates have been observed elsewhere in Asian coastal areas, suggesting that such erosion is widespread.

Our multivariate analysis examining possible factors underlying observed coastal changes was limited by a lack of long-term oceanographic data and the spatial resolution of available quantitative data (province as a basis). Nonetheless, we were able to quantify aspects of terrestrial hydrology, coastal morphology, coastal land use, and mangrove dynamics. It was found that net coastal area loss, i.e. accretion minus erosion, was governed by fetch, coastal length, catchment area and the presence of dams. Coasts with longer fetch are more strongly attacked by waves during monsoonal periods and thus are probably more vulnerable to loose coastal land. A larger catchment area probably also increased exposure to the monsoons. More importantly, the presence of dams in the river appeared to aggravate this pattern (Figure 5.7a), probably because it reduced fluvial sediment supply to the littoral zone (Milliman 2001; ISME/GLOMIS 2002; Bonora et al. 2002; Batalla 2003; Bird et al. 2004). Gross erosion was correlated not only to fetch, but also to the increase of shrimp farm area and associated loss in mangrove area. Likewise, this pattern was found to be enhanced by the presence of dams (Figure 5.7b). Unexpectedly, positive accretion also occurred in areas with more shrimp farms and higher fetch to the monsoons. This is probably related to the coincidence of concentrated shrimp farm developments on the wide gradual slope of the eastern coastal plains that have sandbar-dominated coastlines. Discontinued occurrence of erosion and accretion coupled with sand movement may also account for this pattern. The eroded sediment is transported from higher energy segments by littoral drift to lower energy segments and accumulates there. The eastern coast, therefore clearly features as more dynamic than the western coast (Figure 5.4): substantial proportions of the coast are subject to both erosion and accretion.

This study revealed that southern Thai coastlines with and without mangroves behave differently. Mangrove-dominated coastal segments exhibited less erosion while non-vegetated segments or former mangrove areas incurred substantial erosion. The dense structure of mangrove root systems possibly helps consolidate the coastal soil, hence the shoreline is more resistant to erosion (Mazda et al. 1997). Furthermore, mangrove roots reduce flow and promote flocculation and sedimentation upon the soil surface, eventually allowing positive accretion (Furukawa & Wolanski 1996; Smoak & Patchneelam 1999), particularly at river mouths or bays on the eastern coast. On the contrary, exposed and unconsolidated soils of non-vegetated and former mangrove land are more prone to erode. Our multivariate dataset provided correlative patterns supporting the significance of mangroves (e.g. Figure 5.5). Clearly, reduced sediment delivery to the coast has reduced the amount of mangrove habitat. These data also support the quantitative contention made elsewhere (Mazda et al. 1997) that mangroves reduce erosion (Figure 5.5b and 5.5c; Figure 5.6).

Our study suggests that apart from natural phenomena such as exposure to wave attack during the monsoons, anthropogenic activities have had severe impacts on changes of southern Thai coastlines, particularly in relation to conversion of mangrove area to shrimp ponds and the damming of major rivers. Therefore, the existence of mangroves and a continued riverine sediment flux are crucial to maintain coastal stability in the region. Indeed, presently due to extensive coastal development and inland damming, erosion prevails at rates around 1.3 m y^{-1}.

Acknowledgements

The authors are grateful to the Netherlands Foundation for the Advancement of Tropical Research: WOTRO (project WB 84-412) for financially supported this study. We would like to thank The Marine Section of The Pollution Control Department for providing provincial sediment data and C. Worachina, W. Rattananond and the forestry staff at Ao Phang-nga National Park for their assistance in the fields. We are also grateful to Prof. Patrick Denny for his suggestions and critically reading of the manuscript.

References

Aksornkoae, S. 1993. Ecology and management of mangroves. IUCN, Bangkok, Thailand, 176 pp.

Batalla, R.J. 2003. Sediment deficit in rivers caused by dams and in stream gravel mining. A review with examples from NE Spain. Cuaternario y Geomorfología 17: 79-91.

Blasco, F., Saenger, P. and Janodet, E. 1996. Mangroves as indicators of coastal change. Catena 27: 167-178.

Bonora, N., Immordino, F, Schiavi, C., Simeoni, U. and Valpreda, E. 2002. Interaction between catchment basin management and coastal evolution (Southern Italy). Journal of Coastal Research 36: 81-88.

Berger, U. and Hildenbrandt, H. 2000. A new approach to spatially explicit modelling of forest dynamics: spacing, aging and neighbourhood competition of mangrove trees. Ecological Modelling 132: 287-302.

Berger, U., Hildenbrandt, H. and Grimm, V. 2002. Towards a standard for the individual-based modeling of plant populations: self-thinning and the field-of-neighborhood approach. Natural Resource Modeling 15: 39-54.

Bird, M., Chua, S., Fifield, L.K., Teh, T.S. and Lai, J. 2004. Evolution of the Sungei Buloh-Kranji mangrove coast, Singapore. Applied Geography 24: 181-198.

Charuppat, T. and Charuppat, J. 1997. Application of Landsat 5-TM for monitoring the changes of mangrove forest area in Thailand. In: Proceeding of the 10th National Seminar on Mangrove Ecology, 25-28 August 1997, Hat Yai, Thailand. National Research Council of Thailand, V.9: 1-8 (in Thai).

Duarte, C.M., Thampanya, U., Terrados, J., Geertz-Hansen, O. and Fortes, M.D. 1999. The determination of the age and growth of SE Asian mangrove seedlings from internodal counts. Mangroves and Salt Marshes 3: 251-257.

Ellison, J.C. 1993. Mangrove retreat with rising sea level, Bermuda. Estuarine, Coastal and Shelf Science 37: 75–87

Elster, C. 2000. Reason for reforestation success and failure with three mangrove species in Colombia. Forest Ecology and Management 131: 201-214.

FAO 1994. Mangrove forest management guidelines. FAO Forestry paper no. 117. FAO, Rome, Italy, 319 pp.

Field, C.D. 1995. Impact of expected climate change on mangroves. Hydrobiologia 295: 75–81.

Field, C.D. 1999. Mangrove rehabilitation: choice and necessity. Hydrobiologia 413: 47-52.

Furukawa, K. and Wolanski, E. 1996. Sedimentation in mangrove forests. Mangroves and Salt Marshes 1 : 3-10.

Gilbert, A.J. and Janssen, R. 1998. Use of environmental functions to communicate the values of a mangrove ecosystem under different management regimes. Ecological Economics 25: 323-346.

Ghosh, T., Bhandari G. and Hazra, S. 2001. Assessment of land use/land cover dynamics and shoreline changes of Sagar Island through remote sensing. In: In: Proceeding of The 22th Asian Conference on Remote Sensing. 5.9, November, 2001. Singapore. Singapore National University, Singapore Institute of Surveyors and Values and Asian Association on Remote Sensing.

Havanond, S. 1995. Re-afforestation of mangrove forests in Thailand. In: Khemnark, C. (ed.), Ecology and Management of Mangrove Restoration and Regeneration in East and Southeast Asia, Proceeding of the ECOTONE IV, pp. 203-216. Kasetsart University, Bangkok, Thailand.

Hogarth, P.J. 2001. Mangroves and global climate change. In: Borgese, E.M., Chircop, A. and McConnell, M. (eds.), Environment and Coastal Management, Ocean Yearbook 15, pp.331-349. The University of Chicago Press.

ISME/GLOMIS. River damming and changes in mangrove distribution. ISME/GLOMIS Electronic Journal. Volume 2, July 2002.

Janssen, R. and Padilla, J.E. 1999. Preservation or conservation? Valuation and evaluation of a mangrove forest in the Philippines. Environmental and Resource Economics 14: 297-331.

Kamp-Nielsen L.,Vermaat J.E.,Wesseling I.,Borum J. and Geertz-Hansen O. 2002. Sediment properties along gradients of siltation in South-east Asia. Estuarine, Coastal and Shelf Science 54 : 127-137.

Khemnark, C. 1995. Ecology and management of mangrove restoration and regeneration in East and Southeast Asia. Proceeding of the ECOTONE IV, 18-22 January 1995, Surat Thani, Thailand. Amarin Co Ltd., Bangkok, Thailand, 339 pp.

Maged, M.M. and Mansor, S.B. 1998. Coastal erosion modeling using remotely sensed data. In: Proceeding of The 19th Asian Conference on Remote Sensing. 16-20, November, 1998. Manila, Philippines. National Mapping and Resource Information Authority and Asian Association of Remote Sensing.

Maged, M.M., Mansor, S.B. and Mohamed, M.I.H. 1998. Shoreline change detection Using TOPSAR/AIRSAR data. In: Proceeding of The 19th Asian Conference on Remote Sensing. 16-20, November, 1998. Manila, Philippines. National Mapping and Resource Information Authority and Asian Association of Remote Sensing

Mazda, Y., Magi, M., Kogo, M. and Hong, P.N. 1997. Mangroves as a coastal protection from waves in the Tong King delta, Vietnam. Mangroves and Salt Marshes 1:127-135.

Meteorological Department. 2004. Rainfall data 1971-2000. The Meteorological Department 4353 Sukhumvit, Bangkok, Thailand. Website: www.tmd.go.th

Milliman, J.D. 2001. Delivery and fate of fluvial water and sediment to the sea: a marine geologist's view of European rivers. Scientia Marina 65: 121-132.

Panapitukkul, N., Duarte, C.M., Thampanya, U., Kheowvongsri, P., Srichai, N., Geertz-Hansen, O., Terrados, J. and Boromthanarath, S. 1998. Mangrove colonization: mangrove progression over the growing Pak Phanang (SE Thailand) mudflat. Estuarine, Coastal and Shelf Science 47: 51-61.

Plathong, J. and Sitthirach, N. 1998. Traditional and current use of mangrove forest in Southern Thailand. Wetlands International-Thailand Programme/PSU, Publication No.3, 91 pp.

Pollution Control Department. 2003. Water quality survey along Southern Thai coastline. Marine Water Division, Pollution Control Department, 92 Phahon Yothin road, Bangkok, Thailand. Website: www.pcd.go.th.

Purba, M. 1991. Impact of high sedimentation rate on the coastal resources of Segara Anakan, Indonesia, p.143-152. In L.M. Chou, T.E. Chua, H.W. Khoo, P.E. Lim, J.N. Paw, G.T. Silvertre, M.J. Valencia, A.T. White and P.K. Wong (eds) Towards an integrated management of tropical coastal resources. ICLARM conference proceedings 22, 445 pp.

Rao, A.N. 1986. Mangrove ecosystems of Asia and the Pacific. In: Mangrove of Asia and the Pacific: Status and Management, pp. 1-48. Tech. Rep.UNDP/UNESCO.

Robertson, A.I., and Alongi D.M. (ed.). 1992. Tropical mangrove ecosystems. Coastal and estuarine series; 41. American Geophysical Union, Washington, D.C., 330 pp.

Royal Forest Department. 2004. Forest statistics 2002. Royal Forest Department 61 Phaholyathin, Ladyao, Bangkok, Thailand. Website: www.forest.go.th.

Royal Irrigation Department. 2004. Thailand Hydrological Data 1979-2000. Hydrological Division, Royal Irrigation Department, Bangkok, Thailand.

Ruitenbeek, H.J. 1994. Modelling economy-ecology linkages in mangroves: Economic evidence for promoting conservation in Bintuni Bay, Indonesia. Ecological Economics 10: 233-247.

Saito, Y. 2001. Deltas in Southeast and East Asia: Their evolution and current problems. In: Mimura, N. and Yokoki, H. (ed.), Global Change and Asia Pacific Coasts. Proceeding of APN/SURVAS/LOICZ Joint Conference on Coastal Impacts of Climate change and Adaptation in the Asia-Pacific Region, APN, Kobe, Japan, November 14-16, 2000, pp.185-191.

Semesi, A.K. 1998. Mangrove management and utilization in Eastern Africa. Ambio 27: 620-626.

Sinsakul, S., Tiyapairach S., Chaimanee, N. and Aramprayoon, B. 1999. Coastal change along the Andaman Sea Coast of Thailand. Geological Survey Division of the Department of Mineral Resources, Thailand. Research Report. 60 pp.

Sinsakul, S., Tiyapairach, S., Chaimanee, N. and Aramprayoon, B. 2002. Coastal change along the Gulf of Thailand coast. Geological Survey Division of the Department of Mineral Resources, Thailand. Research Report. 173 pp.

Siripong, A. 1985. The characteristics of the ties in the gulf of Thailand. In: Proceeding of the 5th National Seminar on Mangrove Ecology, 26-29 July 1985, Phuket, Thailand. National Research Council of Thailand, V.1: 1-15 (in Thai).

Smoak, J.M. and Patchneelam, S.R. 1999. Sediment mixing and accumulation in a mangrove ecosystem: evidence from ^{210}Pb, ^{234}Th and ^{7}Be. Mangroves and Salt Marshes 3: 17-27.

Thampanya, U., Vermaat, J.E. and Duarte, C.M., 2002a. Colonization success of common Thai mangrove species as a function of shelter from water movement. Marine Ecology Progress Series 237: 111-120.

Thampanya U., Vermaat, J.E. and Terrados, J. 2002b. The effect of increasing sediment accretion on the seedlings of three common Thai mangrove species. Aquatic Botany 74: 315-325.

Thampanya U. and Vermaat, J.E. (in prep). Diameter-age relationship for estimating age of common SE Asian mangrove taxa. (Chapter 4).

Thom, B.E. 1982. Mangrove ecology – A geomorphological perspective. In: Clough, B.F. (ed.), Mangrove ecosystems in Australia, pp.3-17. Australian Institute of Marine Scince and Australian National University Press.

Vongvisessomjai, S. Polsi, R., Manotham, C. and Srisaengthong, D. 1996. Coastal erosion in the Gulf of Thailand. In: Milliman, J.D., and Haq, B.U. (eds.), Sea level rise and coastal subsidence, Kluwer Academic Publishers, 131-150.

Woodroffe, C.D. 1992. Mangrove sediments and geomorphology. In: D. Alongi and A. Robertson (ed.), Tropical Mangrove Ecosystem. American Geophysical Union, Coastal and Estuarine Studies, pp. 7-41.

Chapter 6

Predicting mangrove colonization success: development of a simple, integrative, demographic model

Udomluck Thampanya and Jan E. Vermaat

Abstract

A demographic model of mangrove colonization, seedling establishment and subsequent stand growth was developed for three common SE Asian species: *Avicennia*, *Rhizophora* and *Sonneratia* by synthesizing experiment and field data incorporation with data from the literature. The model involved five stages of the mangrove life span (approximately 100 years) and included the effects of water turbulence, herbivory, sedimentation, salinity and drought condition. It was developed using the mudflats of the Pak Phanang Bay (Southern Thailand) as a template. Unlike other existing mangrove models, the present one includes seedling establishment as a critical stage. The model was simulated over a 30-year period. The three species models produced observed colonized areas well and were found to be most sensitive to changes in water turbulence and seedling herbivory. The early colonizer, *Avicennia*, had colonized more area than the other two species. The area colonized after 30 years simulation were 782 ha, 364 ha and 88 ha or colonized at a rate of approximately 26 ha^{-1}, 12 ha^{-1} and 3 ha^{-1} for *Avicennia*, *Sonneratia* and *Rhizophora*, respectively. A range of scenarios was applied to the models covering changes in rainfall, accelerated sea level and inflow river damming. These scenario runs suggest that the three species have less success in colonization if accelerated sea level rise occur. A slightly (10%) lower rainfall enhances colonization of *Avicennia* and *Rhizophora* while 50% more rainfall is favorable only for *Rhizophora*. In case of full river damming, colonization of *Avicennia* increases drastically. In contrast, colonization by *Sonneratia* decreased by more than 80%.

To be submitted to Ecosystems

Introduction

Mangrove tree species manage to develop dense and productive stands at the interface of land and sea (Woodroffe 1992), where conditions are stressful (Hutching & Saenger 1987) and a suite of interacting factors creates complex variability for spatial and temporal scales of habitats (Thom 1984; Clarke & Allaway 1993; Panapitukkul et al. 1998). Individual factors have been studied in the field (Clarke & Allaway 1993; Patterson et al. 1997; Terrados et al. 1997; Elster et al. 1999; Clake & Kerrigan 2002; Thampanya et al. 2002a; 2002b; Clarke 2004) as well as in laboratories (McMillan 1971; Patterson et al. 1997; Ellison & Farnsworth 1997; Aziz & Khan 2001). For a better understanding of such a complex interplay, simultaneous integration of these factors is often favourably covered in a mathematical model (Chen & Twilley 1998; Berger & Hildenbrandt 2000; Berger et al. 2002). Such models often start at the established sapling stage of the life cycle, possibly due to limitation of site-specific available field data.

The most critical period of the mangrove life cycle is probably, however, the colonization or establishment phase when propagules take root on unoccupied substrate (Duke 2001). The success of establishment is generally governed by a complex interaction of regulating factors such as water current (Clarke 1993; Thampanya et al. 2002a) and suitability of substrate (Clarke 1993; McKee 1995; Lee et al. 1996; Bosire et al. 2003) with the demographic and life cycle characteristics of different taxa. The published demographic models including this early stage of mangroves are still rare (Clarke 1995; Chen & Twilley 1998; Delgado et al. 1999).

This study attempts to develop demographic models for three common SE Asia mangrove taxa: *Avicennia alba, Rhizophora apiculata* and *Sonneratia caseolaris** since these tree species usually colonize the outer edge of mangrove forest (WWF website 2005; Thampanya pers. Observ.). We synthesized our own experimental and field data (Terrados et al. 1997; Panapitukkul et al. 1998; Thampanya et al. 2002a, Chapter 2; Thampanya et al. 2002b, Chapter 3) with those available from literature (Rabinowitz 1978; Putz & Chan 1986; Smith et al. 1989; Floss 1993). The aim of these models is to obtain an overall perspective of colonization success on unoccupied intertidal mudflats in relation to the most important environmental factors. The success in colonization was elucidated in a five stage model starting from seedling establishment up to the adult tree stage. We examined rainfall, water turbulence, sedimentation, drought and herbivory as important factors influencing seedling establishment on open mudflats. Salinity is also taken into account, although it was reported to unlikely affect establishment (McMillan 1971; Clarke 1995). The resultant models are expected to provide useful tools to improve our integrated understanding of mangrove existence vis-a-vis the complexity of coastal dynamics and support coastal resource management on mangrove reforestation schemes.

Type of area

Potential habitats for mangroves to colonize are open, newly-formed mudflats in the vicinity of existing mangrove forests. In our study, we used the exposed mudflats in Pak Phanang Bay, Nakhon Si Thammarat province, Southern Thailand as a template (Figure 2.1, Chapter 2). The area has a tropical monsoon climate type with annual precipitation of about 2,035-2,750 mm and over 170 rainy days per year. Daily mean air temperature varies between 28-33 °C (Meteorological Department). The environmental setting of the case study area can be classified as river-dominated (Woodroffe 1992). This semi-enclosed

*hereafter, we use genus names to facilitate reading

bay receives continuous fresh water discharge from Pak Phanang River and surrounding canals along with sea water from tidal flooding (Figure 2.1, Chapter 2). Thus, water in the bay varies between brackish and sea water. Water salinity fluctuates between 2-14 ppt during the rainy season and 8-33 ppt during the dry season (Figure 2.2, Chapter 2). Tides are mixed (diurnal) and the maximum amplitudes are approximately 0.70 metre for neap tide and 1.1 metre for spring tide (JICA 1987). Regardless direction, water current in the main channels ranges from 0.2 to 0.7 meter per second (JICA 1987). The bay receives approximately 2×10^5 tons of riverine sediments annually (Flos 1993). The forest is inundated approximately 20 times monthly while for mudflats this is approximately 45 times. The mudflats are partly exposed at low tide and are covered with at least 50 cm of water at high tide. The mudflat soil consists of 30-40% silt and 50% clay, 3-10% organic matter and has a water content of 22-50% while nutrient availablity is quite homogeneous with 30-40 mg kg^{-1} of available P and total N ranges between 0.10-0.15% (Thampanya unpubl. data). During the fruit falling period of *Avicennia* and *Rhizophora*, which is usually coincident with the rainy season, the area receives massive amounts of mangrove propagules flushed from the forest.

Model conceptualization

The mangrove colonization model focuses on establishment success in early growth stages and recruitment to the next stages of the mangrove life span (estimated to be approximately 100 years; Verheyden et al. 2004) i.e. from propagule to seedling, sapling and tree. An overview of the conceptual model is illustrated in Figure 6.1. The model commences when seeds (or seedlings in case of *Avicennia* and *Rhizophora*) are abscised from mature trees (and considered as propagule thereafter). Being inundated, these buoyant propagules are dispersed by water currents. The obligate dispersal period varies with local hydrodynamics and species, for example, one week for *Avicennia marina* in Australia (Clarke 1993; Clarke et al. 2001), two weeks for *Avicennia germinans* and 40 days for *Rhizophora mangle* in Panama (Rabinowitz 1978). Some of these propagules may succeed in establishment while others will be lost to the open sea. The chance of propagules to establish themselves on the mudflats comes when they are able to reach the substrate because of low tide or after loosing buoyancy. Water turbulence, however, has a considerable effect on propagule dispersal and may hinder the establishment (Clarke 1993; Thampanya et al. 2002a), whilst strong currents are also harmful to established seedlings. Terrestrial runoff, wind and irregular storms are important factors driving water turbulence, sedimentation, salinity and nutrient availability. This run-off is highly related to patterns of rainfall and anthropogenic activities (i.e. land use). Herbivory is another factor that causes failure in propagule establishment and seedling mortality (Patterson et al. 1997; Clarke & Kerrigan 2002), similar to sedimentation (Terrados et al. 1997; Thampanya et al. 2002b). The anchored propagules can be attacked easily by crabs (Patterson et al. 1997; Clarke & Kerrigan 2002) or be dispersed again by water motion before a firm rooting is accomplished. Local environmental conditions such as light intensity, salinity and nutrient availability are considered as factors influencing survival and growth after a seedling has established on the substrate. During recruitment to the next stages, mortality is common (Clarke 1995). Storm surges with a low frequency can lead to sudden and large–scale death of saplings and trees. The cycle is completed when the tree reaches maturity and is able to flower and produce seeds.

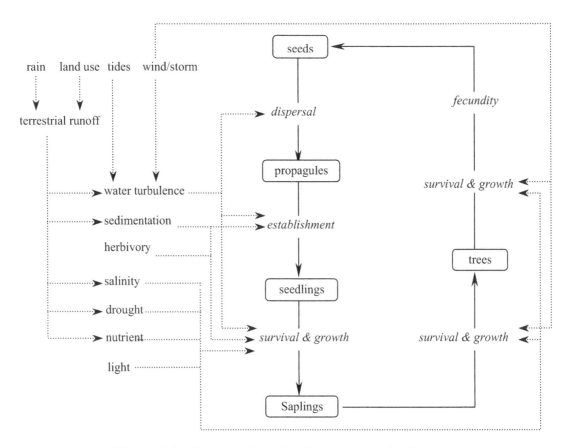

Figure 6.1. Conceptual model of mangrove colonization

Model boundary

The predictive models have been developed in accordance with the mangrove life cycle conceptualized in Figure 6.1. Successful colonization follows from propagule dispersal, seedling establishment and, finally, recruitment to sapling and tree. Since the modelling focuses on the effects of water turbulence and sedimentation, local environmental factors such as nutrient and light availability are assumed to be in acceptable ranges during establishment and incorporated in recruitment to the next stages. Land use pattern, wind and storm surges are beyond the scope of this study. For each species, a separate model was developed and the simulation has been executed independently. Colonization success will be examined using number of colonized trees or colonized area at the end of simulation time.

Data and parameterization

Data necessary for model development were obtained from various sources. Demographic data of young stages are from our own field observations and experiments *i.e.*, available propagules, seed or seedling production, plant density, seedling establishment and mortality (Terrados et al. 1997 and Thampanya et al. 2002a; 2002b). Germination, herbivory and demographic data of sapling and tree are obtained from literature (Wechakit 1987; Bamroongrugsa 1997; Chukwamdee & Anunsiriwat 1997; Chen et. al 2000). Local conditions such as rainfall, discharge, salinity, drought and sediment load were obtained from field observation (Thampanya et al. 2002a; 2002b) as well as secondary data from reports and literature (JICA 1987; Flos 1993; Meteorological Department 2004). Seedling data were divided into two stages: young seedling (seedlings ≤ 3 month old) and old seedling (seedlings >3 month) since the first three months of the life cycle is probably the most critical and mortality was reported to be highest (Elster 2000).

Propagule flux estimation

In Pak Phanang Bay, fruits of *Avicennia alba* and *Rhizophora apiculata* fall during the rainy season (November-January, Chapter 1). These fruits (propagules) are dispersed by water current through the canals discharging into the bay. Quantification of available propagules was made possible from daily number of propagules trapped by stationary fishing nets located at the mouth of these canals. Four main canals were monitored during the rainy season of 1997-1999 with the assistance of local fishermen. An estimation of arriving propagules to the forest edge of these two species was made using the relationship between daily number of trapped propagules and local daily rainfall. The regression equation with a five-days running average is applied (Figure 6.2).

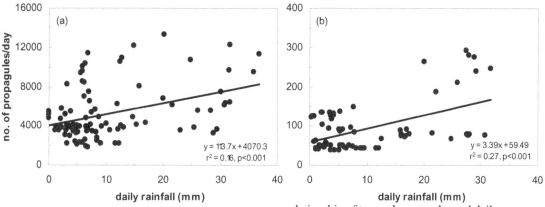

Figure 6.2. Daily arriving propagules estimated by the relationship of trapped propagules and daily rainfall: (a) *Avicennia alba* and (b) *Rhizophora apiculata*.

Fruits of *Sonneratia caseolaris* fall after the rainy season (February-March, Chapter 1), unfortunately, data on arriving propagules were not available. Nevertheless, during our field observations we recorded number of fruits produced per tree for 60 individual trees and the percentage of fruiting trees in the patches. Hence, fruit production of this species can be estimated from average number of fruit produced per tree and an estimated number of maternal trees at the forest edge. In addition, the number of seed per fruit was counted for three selected fruits (approximately 500, 800 and 1100, respectively), however, the minimum number (500) was used in model simulations to avoid calculation overflow.

Effect of water turbulence, salinity and sedimentation

Seedling loss to the open sea by water movement was estimated from field observations (Thampanya unpubl. data, Figure 6.3a). Seedling germination (Figure 6.3b) of *A. alba* and *S. caseolaris* and survival affected by salinity are obtained from six-month experiments in greenhouses (Patanaponpaiboon & Paliyavuth 2001; Jirawattanapun et al. 2002), whilst the *R. apiculata* experiment was carried out for eight months in a nursery near Pattani Bay,

Southern Thailand (Bamroongrugsa & Kaewwongsri 2000). Seedlings Survival in relation
to salinity for young seedling is depicted in Figure 6.4a and old seedling in Figure 6.4b.

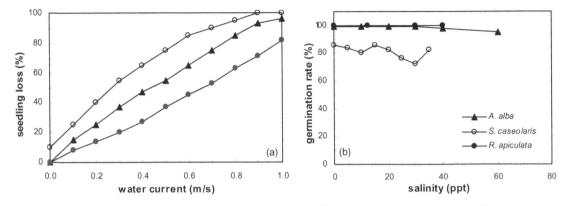

Figure 6.3. (a) Seedling loss due to water turbulence and (b) seedling germination in relation to salinity,
n=150 for *A. alba* and *S. caseolaris* and n=200 for *R. apiculata*.

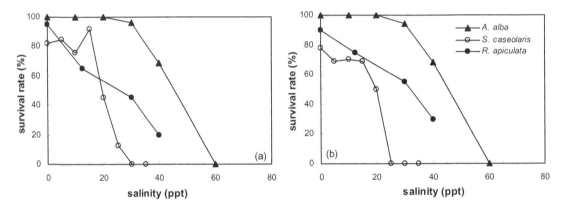

Figure 6.4. Seedling survival in relation to salinity for (a) young seedlings (≤ 3 months) and (b) old seedlings
(>3 months), n=150 for *A. alba* and *S. caseolaris* and n=200 for *R. apiculata*.

Sediment accretion of the study area is estimated from the relationship between historical
freshwater runoff and sediment loads (Flos, 1993; Figure 6.5a). Water current intensity
was approximated from the relationship between daily rainfall and freshwater flow, since
the tidal amplitude is small and has little effect (Figure 6.5b). Data necessary to model
simulation are summarized in Table 6.1 and 6.2. Data on seedling mortality caused by
water turbulence followed Thampanya et al. (2002a). Mortality due to sediment burial for
Rhizophora apiculata followed Terrados et al. (1997), whilst for *Avicennia alba* and
Sonneratia caseolaris followed Thampanya et al. (2002b). Mortality equations used in the
model for the three species are as followed:

> *Avicennia alba* mortality   = 0.001 * sediment level * burial time
> *Rhizophora apiculata* mortality = 0.001 * sediment level * burial time
> *Sonneratia caseolaris* mortality = 0.0002 * sediment level * burial time

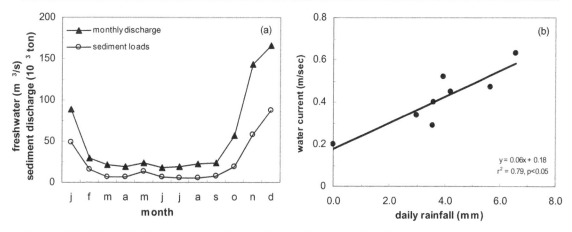

Figure 6.5. (a) Monthly freshwater runoff and sediment discharge of Pak Phanang Bay (average from a period of 1960-1969) and (b) water current estimated from rainfall (data source: Flos 1993).

Table 6.1. Mangrove data used in the model.

Data	species			unit	source
	A. alba	R. apiculata	S. caseolaris		
mortality caused by herbivory					approx. from Smith 1989; Clarke & Kerrigan 2002
- young seedling	60	30	70	%	
- old seedling	55	25	60		
mortality due to drought				%	
- young seedling	5	3	5		
- old seedling	3	2	4		user defined
- sapling	0	0	0.5		
- tree	0	0	0		
tree mortality (by unknown factor)	3	3	3	%	Putz&Chan 1986; Delgado et al 1999; Wechakit 1987
average fruit production per tree					
- sapling stage	300	20	50	fruits/tree	pers. obs.
- tree stage	2000	200	150		
percentage of fruiting					
- sapling stage	50	50	50	%	pers. obs.
- tree stage	60	60	60		
propagule retention time	5	7	10	days	pers. obs.
recruit time					
- from seedling to sapling	4	4	3	years	cf. Chapter 4
- from sapling to tree	6	6	5		
approximate tree density after colonization	2,110 ±130	1,750 ±67	1,570 ±136	individuals /ha	Kaewwongsri & Bamrungragsa 1997
arriving propagules generated from newly colonized area after plants reach maturity	15	3	75	%	approximated from field data
seed decaying time	-	-	10	days	pers. obs.
number of seed per fruit	1	1	500	seeds	pers. obs..Chapter 1
number of maternal trees	-	-	25,000	trees/area	pers. obs.

Table 6.2. Environmental data used in the model.

data	value	unit	source
catchment area	1, 070	km^2	JICA 1987
sediment deposit area flooded by river	125	km^2	JICA 1987
area of the Pak Phanang Bay	80	km^2	Flos 1993
sW (specific weight per volume of mud)	1.3 x 10^3	kg/m^3	JICA 1987
runoff coefficient	55	%	Flos 1993
average rainfall (1960-1990)	0-90	mm d^{-1}	Meteorological Department 2004
salinity	0-40	ppt	approx. from field data

Model Formulation

Model formulation and testing has been accomplished with the system dynamics modelling tool, Stella (HPS, 1997; Costanza et al. 1998). The formulation starts with identifying system variables, then connecting all identified variables together according to their relationships. These connected variables are composed as a flow diagram using graphical symbols provided in Stella (HPS, 1997). Stella applies three common types of system variables: stock, converter and flow. For instance, propagules, seedlings, saplings and trees are stock variables since they accumulate elements and their population sizes will be monitored. Herbivory, rainfall, runoff, sedimentation, burial and water turbulence are taken here as converter variables because they contain constants or equations. Establishment, growth, mortality and recruitment are flow variables because they transmit elements from one stock to another or to system outflow controlled by associated converters. One important variable namely 'annual cycle' has been incorporated and assigned to act as system stimulator and annual controller necessary for model running.

Figure 6.6 depicts the flow diagram of mangrove colonization for *Avicennia alba* and *Rhizophora apiculata* as they have similar demographic patterns. The part inside the box illustrates the mangrove demographic process, whilst that outside represents the process of environmental factors influencing seedling colonization, i.e., water turbulence and sedimentation. Both processes (demographic and environmental factor) have the same starting point (the annual cycle on the top right corner) and run concurrently. The colonization process starts with available propagules arrival to the area, some are lost and flowed out of the system due to water current but some are able to germinate and establish on the mud. The main causes of mortality in the seedling stage (both young and old seedlings) are assumed to be herbivory, drought, water turbulence, salinity and sediment burial (Clarke & Allaway 1993; Elster et al. 1999; Ellison 1999; Thampanya et al. 2002a; 2002b; Clarke & Kerrigan 2002). For the later stages, however, due to lack of demographic data as well as for simplicity, mortality causes are assumed to be sediment burial and drought at the sapling stage, drought and an additional unknown factor for the adult tree stage. Once the colonized plants reach maturity (sapling and tree stages), they generate additional propagules to the system.

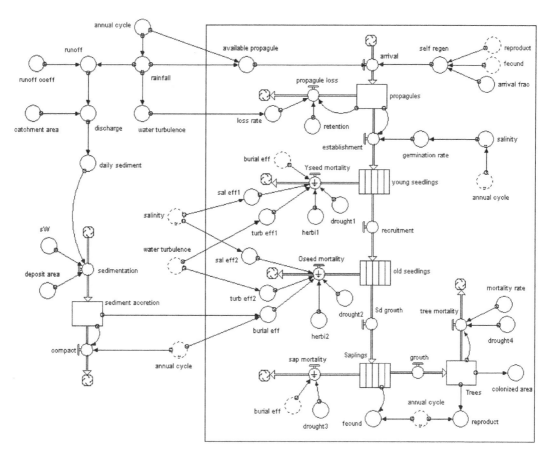

Figure 6.6. Flow diagram of colonization for *Avicennia alba* and *Rhizopora apiculata*.

For *Sonneratia caseolaris*, the colonization is more complex (Figure 6.7). The propagules of this species arrive in a fruit containing many seeds (Table 6.1). After sinking and anchorage in the mud, the fruit needs around 10 days to decay before its sticky seeds emerge and germinate. Thereafter, the modelled pattern is similar to that of the two other species. Among environmental processes, rainfall is the main variable that controls water turbulence since hydrodynamics in the area is river-dominated and the tidal amplitude is quite small (less than 0.5 m, JICA 1987). The rainfall is associated with catchment area and a runoff coefficient to generate discharge and sedimentation.

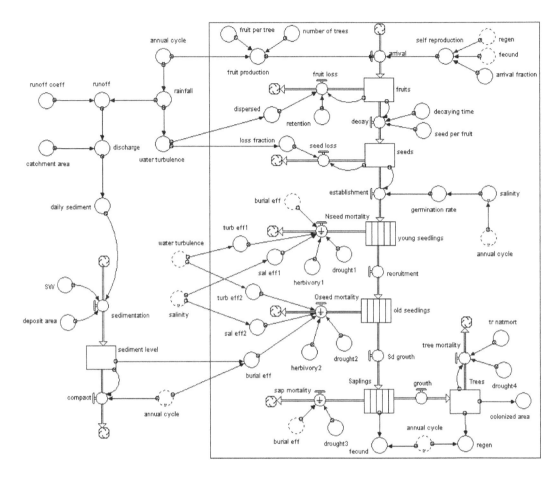

Figure 6.7. Flow diagram of colonization for *Sonneratia caseolaris.*

Simulation results

The model of each species is simulated for 30 years, which should be a sufficiently long period to reveal differences in colonization success. The simulation results show drastic fluctuations in seedling populations of all species whilst fluctuations in the numbers of saplings and trees are also substantial (Figure 6.8). When applying the approximate colonized tree density estimated in Table 6.1, *Avicennia* had colonized 782 ha at the end of the simulation, or about 26.1 ha y^{-1}. Its colonization gradually increases for the first 20 years but thereafter, the increase is quite steep (Figure 6.9). For *Sonneratia*, the colonization appeared earlier and increases gradually through the simulation time. At the end of the simulation, it had colonized 364 ha or at a rate of 12.1 ha y^{-1}. *Rhizophora* seems to have less success in colonization. It colonized 88 ha within 30 years or at a rate of 2.9 ha y^{-1}. Seedlings of *Avicennia* arrived to the area in higher numbers than *Rhizophora* and *Sonneratia*. Accordingly, the number of colonized trees of this species was also the highest (1,650,920 individuals) compared to 156,749 and 572,201 individuals for *Rhizophora* and *Sonneratia* or higher than the latter two species by eleven and three times, respectively.

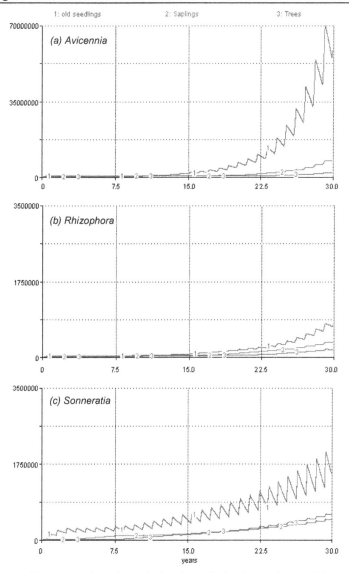

Figure 6.8. Fluctuation of population through simulation time: (a) *Avicennia alba* (b) *Rhizophora apiculata* and (c) *Sonneratia caseolaris*.

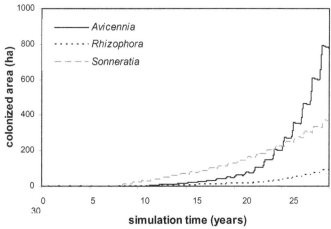

Figure 6.9. Estimated colonized area of the three species through simulation period. Note that the total area of Pak Phanang Bay (the study area) is 8,000 ha.

Model Sensitivity Analysis

Sensitivity analysis is important for modelling as it examines which input parameters have the strongest impact on the model results. In this study, the model sensitivity analysis has been conducted for four selected parameters, i.e. water turbulence, sedimentation, herbivory and salinity by varying one parameter at a time. Each parameter was first decreased by 10% and 50%, then increased by 10% and 50%. The models of all three species show a very high sensitivity to herbivory and water turbulence. Change in colonization increased drastically when these two parameters were reduced by 50%. In contrast, it became zero, i.e. no colonization was observed (since none of the seedlings survived) when they were increased by 10% and 50% (Figure 6.10a and 6.10c). Sensitivity to sedimentation and salinity was observed to be comparatively small, and varied more or less linearly with percentage change (Figure 6.10b and 6.10d). Furthermore, when amplitudes in responsiveness are compared. *Sonneratia* was found to be most sensitive to changes in turbulence, herbivory and salinity, but *Rhizophora* and *Avicennia* were more sensitive to changes in sedimentation.

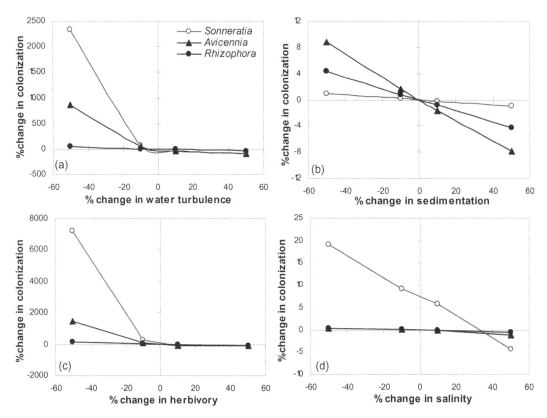

Figure 6.10. Graphical results of model sensitivity to: (a) water turbulence, (b) herbivory, (c) sedimentation, and (d) salinity, for the three species.

Model validation

For validation of the results, modelled colonized areas have been compared with the observed progression of mangrove forest obtained from remote sensing data analysis. For *Avicennia,* the colonization rate resulting from the model simulation is 26.1 ha y^{-1}, while the progress measured of this species which evidently occupied the forest margin of the northern part of the Pak Phanang Bay (unpubl. data Thampanya, Figure 5.2c, Chapter 5) is approximately 755 ha over 31 years (1966-1997) or approximately 24.4 ha y^{-1}. *Sonneratia* mostly occupied the southern part of the Pak Phanang mangrove forest edge and the progression rate resulting from the model simulation was found to be slightly higher than that obtained from the remotely sensed data (12.1 ha y^{-1} vs. 10.0 ha y^{-1}). Unfortunately, the progression of *Rhizophora* in Pak Phanang Bay could not be quantified. Therefore, we compare the result from Pattani Bay instead (Figure 5.2d, Chapter 5). The mangrove forest at Pattani Bay contains mostly mono-specific stands of *Rhizophora apiculata*. The progress measured within 31 years (1967-1998) from remote sensing method of this site is approximately 150 ha or progress at a rate of 4.8 ha y^{-1} which is higher than the result from our model simulation (2.9 ha y^{-1}).

For a more detailed validation, we combined the predictive results of a normal situation with the results of the three mildest scenarios: 10% lower rain fall, rapid sea level rise and 10% lower rain fall with sea level rise (see next paragraph) to get an order of magnitude estimate of potential annual variability. The t-test shows that colonization rates of *Avicennia* and *Sonneratia* obtained from model simulations are not significant different from those resulting from remote sensing analysis, whilst for *Rhizophora*, it was significantly different (Table 6.3). The latter is probably because of site differences. In addition, the adult tree density values used in our model (Table 6.1) are comparable to those obtained from the literature for similar species of *Avicennia* and *Rhizophora* (Chen & Twilley 1998; Delgado et al. 1999; Sherman et al. 2000; Bedin 2001; Kairo et al. 2002).

Table 6.3. Comparison of observed and predicted colonization rates for the three mangrove species studied.

species	observed colonization rate (ha^{-1})	predictive (mean ±SE)	P(t-statistic)
Avicennia	24.4	20.7 ±3.9	0.423
Rhizophora	4.8	2.6 ±0.2	0.001
Sonneratia	10.0	8.4±1.7	0.417

Model prediction: application of scenarios

To predict mangrove colonization success under changes in environmental conditions, the model of each species was run under a range of seven scenarios (Table 6.4): double rainfall, 10% lower rainfall, 50% lower rainfall, accelerated sea level rise (30-50 cm by the year 2050) with normal rainfall, accelerated sea level rise with 10% lower rainfall and river damming, whilst normal rainfall (contemporary situation) was used as the base case. We applied these scenarios because sea level rise as well as less and more irregular rainfall have been recognized recently as important phenomena resulting from global climate change, and their impacts have become more noticeable (IPCC 1998; Saito 2001). Likewise, river damming is a common solution for freshwater deficit in many countries. These dams obstruct freshwater as well as sediment discharge to coastal areas (cf. Chapter 5). All these scenarios ultimately affect mangrove communities that inhabit the areas between land and sea (Ellison 1993; Furukawa & Baba 1997; ISME/GLOMIS 2002). In each model simulation, values of environmental parameters related to scenario assumptions were varied accordingly (Table 6.4) except herbivory (the most sensitive parameter) and sedimentation since these parameter have already been incorporated in the rainfall-runoff equation. For example, the double rainfall scenario creates more freshwater discharge into the bay that leads to decline in water salinity and decline in drought condition. More freshwater discharge also enhances water turbulence leading to increase propagule transportation from the forest but reduce propagule retention time. In contrast, 10% and 50% lower rainfall scenarios provide less freshwater to the area that leads to increase in water salinity by 5% and 25% as well as increase drought condition by 10% and 50%, respectively. Less river discharge also diminishes water turbulence and arrival of propagules whilst it provides more propagule retention time. Sea level rise induces saline water intrusion, increase water depth and increase effect of wave (IPCC 1998). These situations enhance water turbulence and increase salinity (Hull & Titus 1986; Furukawa & Baba 1997). In the full river-damming scenario, freshwater discharge is completely cut off causing a decrease in water turbulence and exportation of mangrove seedlings to the mudflat. Lack of freshwater discharge increases seed retention time as well as drought conditions. Scenario runs were calculated for 30 years as before, and are compared to a "normal", present-day-conditions scenario, for which a band of variability was estimated using the spatial and interannual variability observed in Panapitukkul et al. 1998.

Table 6.4. Changes of relevant parameters used for each scenario (see text for explanation).

scenario	description	salinity	turbulence	drought	seed arrival fraction	seed retention time
NR	normal rainfall	normal	normal	normal	normal	normal
DR	double rainfall	-25%	+50%	-50%	+50%	-50%
10LR	10% lower rainfall	+5%	-10%	+10%	-10%	+10%
50LR	50% lower rain	+25%	-25%	+50%	-50%	+50%
NRSLR	SLR with normal rainfall	+25%	+10%	normal	normal	-10%
10LSLR	SLR with 10% lower rainfall	+30%	+5%	+10%	-10%	+10%
DAM	full river damming	sea water	-50%	+50%	-50%	+80%

The three species responded differently to the scenario conditions. *Avicennia* showed less colonization success than the base case under four scenarios: double rainfall, 50% lower rainfall and accelerated sea level rise with either normal rainfall or 10% lower rainfall (Figure 6.10a). This species, however, was able to colonize more area than in the base case in scenarios with river damming. *Rhizophora* had more colonization success in the double rainfall scenarios whilst 50% lower rainfall, accelerated sea level rise and river damming reduced colonization success (Figure 6.10b). For *Sonneratia*, the colonization success appeared to have decreased in all scenarios, especially in double rainfall and river damming (more than 80%, Figure 6.11c). *Sonneratia*, thus, appeared the most sensitive to changes in the freshwater/seawater balance and sea level rise. Only *Avicennia* responded positively to reduced freshwater inflow. This is in agreement with the patterns in sensitivity observed above (Figure 6.10)

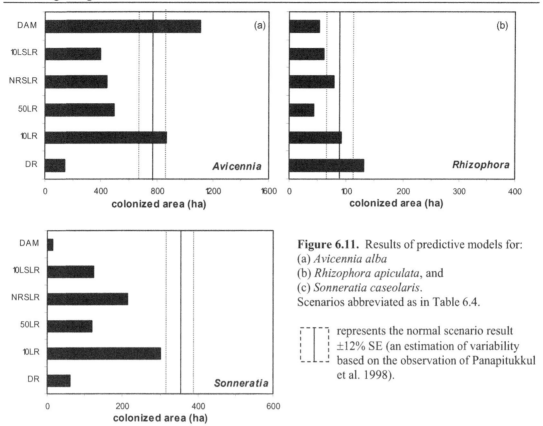

Figure 6.11. Results of predictive models for:
(a) *Avicennia alba*
(b) *Rhizophora apiculata*, and
(c) *Sonneratia caseolaris*.
Scenarios abbreviated as in Table 6.4.

represents the normal scenario result ±12% SE (an estimation of variability based on the observation of Panapitukkul et al. 1998).

Discussion

The three models of early mangrove colonization demography confirm observed patterns: the early colonizer *Avicennia* indeed produced the highest number of propagules and had the highest success in colonization on open substrates. Model validation suggested that simulated colonization rates of the three species within 30 years were comparable to those observed both in Pak Phanang Bay and Pattani Bay (Table 6.3).

Our analyses demonstrated the highest sensitivity of colonization to water turbulence and herbivory (Figure 6.10). Turbulence variability was indeed suggested to be responsible for differences in colonization success across the intertidal landscape and particularly the strong monsoon-related currents were observed to remove large proportions of recently established seedlings (Thampanya et al. 2002a). Herbivory data used in this study, however, were estimated from the literature (Smith 1989; Clarke & Kerrigan 2002), and direct quantitative observations for Southern Thailand were not available. This parameter, therefore, probably has been modelled with the highest uncertainty. Its strong influence in the sensitivity test suggests that confirmation from the field is needed. Elsewhere in Thailand, indeed crab herbivory was found to be an obstacle to seedling survival (Havanond & Maxwell 1993). Our own field observations (Thampanya, pers. observ.) suggest that sensitivity to the above factors diminishes greatly once the sapling stage is reached, simply due to the large size of the plant. Our modelling results indeed confirm the critical sensitivity of the seedling establishment phase. Since the area available for colonization in this study (Pak Phanang Bay) was much larger than that colonized in the 30 years period, it can be assumed that colonization would continue beyond 30 years in a semi-steady-state pattern as suggested by the curves in figure 6.8. The 30 year time horizon was enforced by computational limitations due to the small time step used (a day).

The climate change scenario applications reveal both success and failure in colonization of the three species studied (Figure 6.11). All three species display less success in colonization if accelerated sea level rise takes place. The reason may be that mortality increases due to higher salinity and turbulence as well as longer inundation periods. This is in agreement with the expected retreat in mangrove forests if rapid sea level rise take place (Ellison 1993; Blasco et al. 1996; Yulianto et al. 2004). In the double rainfall scenario, *Avicennia* and *Sonneratia* were able to colonize 80% less area than under normal rainfall. *Rhizophora*, however, was able to colonize more area than in the base case. This is probably because more freshwater discharge not only enhances exportation of propagules from under the canopy onto the mudflats but also increases dispersal and subsequent loss of the smaller propagules of *Avicennia* and *Sonneratia* to the open sea. In addition, increasing water turbulence also hinders propagule establishment (cf. Thampanya et al. 2002a). On the contrary, slightly lower rainfall appeared preferable for establishment, particularly for *Avicennia* and *Rhizophora* seedlings. For *Sonneratia*, the optimal salinity for growth is less than 20 ppt, therefore, lack of freshwater is harmful to their establishment and survival (Figure 6.4; Ball & Pidsley 1995). In the case of fresh water cutoff, the only species that can survive and colonize is *Avicennia*. This is probably because more time is available for propagules to remain in the area and succeed in establishment and because *Avicennia* is well recognized as a salinity-tolerant taxon (Figure 6.4; Clarke & Allaway 1993; Chen & Twilley 1998).

The simple demographic modelling in this study has improved our integrated understanding of mangrove colonization success on open habitats. The three species exhibit similar colonization behavior but have different levels of responses to environmental factors as well as different magnitudes of colonization success. The modelling revealed that besides water turbulence and sedimentation factors, there are other factors that may have severe impacts on colonization such as herbivory and the arrival of propagules to the site. Therefore, in sound planning of mangrove rehabilitation, more ecologically relevant factors should be taken into account. This understanding will enhance the predictability and success of management strategies, such as afforestation programs for coastal protection (Mazda et al. 1997) or enhancement of fishery habitats (Kaly & Jones 1998). We have suggested previously (Chapter 2) that assisting natural mangrove progression in suitable habitats may be an economical alternative for afforestation schemes provided natural fluxes of propagule arrive and are retained in the site.

Acknowledgements

We thank the Netherlands Foundation for the Advancement of Tropical Research (WOTRO) for financially supported this study. The authors are grateful to Dr. Sakda Jongkaewwattana at Chiang Mai University and Johan van de Koppel presently at Netherlands Institute of Ecology (NIOO, Yerseke) and Florian Eppink at the Institute for Environmental Studies of the Vrije Universiteit Amsterdam (IVM-VU) for advice in modelling. We are also grateful to Prof. Patrick Denny for his suggestions in the manuscript.

References

Aziz, I. and Khan. M.A. 2001. Experimental assessment of salinity tolerance of *Ceriops tagal* seedlings and saplings from the Indus Delta, Pakistan. Aquatic Botany 20: 259-268.

Ball, M.C. and Pidsley, S.M. 1995. Growth responses to salinity in relation to distribution of two mangrove species, *Sonneratia alba* and *Sonneratia lanceolata*, in northern Australia. Functional Ecology 9: 77-85.

Bamroongrugsa, N. 1997. Mangrove afforestation on newly accreted mudflats of the Pak Phanang Bay, Nakhon Si Thammarat province. In: Proceeding of the 10th National Seminar on Mangrove Ecology, 25-28 August 1997, Hat Yai, Thailand. National Research Council of Thailand, II3: p1-9 (in Thai).

Bamroongrugsa, N. and Kaewwongsir, P. 2000. A study on watering methods for mangrove seedlings grown in nursery. In: Proceeding of Regional Seminar for East and Southeast Asian Countries: ECOTONE VIII, 23-28 May 1999, Ranong and Phuket, Thailand,   p174-179 (in Thai).

Bedin, T. 2001. The progression of a mangrove forest over a newly formed delta in the Umhlatuze Estuary, South Africa. South African Journal of Botany 67: 433-438.

Blasco, F., Saenger, P. and Janodet, E. 1996. Mangroves as indicators of coastal change. Catena 27: 167-178.

Bosire, J.O., Dahdouh-Guebas, F., Kairo, J.G. and Koedam, N. 2003. Colonization of non-planted mangrove species into restored mangrove stands in Gazi Bay, Kenya. Aquatic Botany 76: 267-279.

Berger, U. and Hildenbrandt, H. 2000. A new approach to spatially explicit modelling of forest dynamics: spacing, aging and neighbourhood competition of mangrove trees. Ecological Modelling 132: 287-302.

Berger, U., Hildenbrandt, H. and Grimm, V. 2002. Towards a standard for the individual-based modeling of plant populations: self-thinning and the field-of-neighborhood approach. Natural Resource Modeling 15: 39-54.

Chen, R. and Twilley, R.R. 1998. A gap dynamic model of mangrove forest development along gradients of soil salinity and nutrient resources. Journal of Ecology 86: 37-51.

Chen, Z., Wang, R., Miao, Z. and Gao Y. 2000. Sonneratia reafforestation-restoration of mangroves in China. In Proceeding of Regional Siminar for East and Southeast Asian Countries: ECOTON VIII "Enhancing coastal ecosystem restoration for the 21th century", 23-28 May 1999, Ranong and Phuket, Thailand, p 90-92.

Chukwamdee, J. and Anunsiriwat, A. 1997. Biomass estimation for *Avicennia alba* at Changwat Samut Songkram. In Proceeding of the 10th Thailand National Seminar on Mangrove Ecology, 25-28 August 1997, Hat Yai, Thailand. National Research Council of Thailand, V14: 1-8 (in Thai).

Clarke, P.J. 1993. Dispersal of grey mangrove (*Avicennia marina*) propagules in southeastern Australia. Aquatic Botany 45: 195-204.

Clarke, P.J. 1995. The population dynamics of the mangrove *Avicennia marina*; demographic synthesis and predictive modelling. Hydrobiologia 295: 83-88.

Clarke, P.J. and Allaway, W.G. 1993. The regeneration niche of the grey mangrove (*Avicennia marina*): effects of salinity, light and sediment factors on establishment, growth and survival in the field. Oecologia 93: 548-556.

Clarke, P.J., Kerrigan, R.A. and Westphal, C.J. 2001. Dispersal potential and early growth in 14 tropical mangroves: do early life history traits correlate with patterns of adult distribution? Journal of Ecology 89: 648-659.

Clarke, P.J. and Kerrigan, R.A. 2002. The effects of seed predators on the recruitment of mangroves. Journal of Ecology 90: 728-736.

Clarke, P.J. 2004. Effect of experimental canopy gaps on mangrove recruitment: Lack of habitat partitioning may explain stand dominance. Journal of Ecology 92: 203-213.

Costanza, R., Duplisea, D. and Kautsky, U. 1998. Ecological modelling on modelling ecological and economic systems with STELLA. Ecological Modelling 110: 1-4.

Delgado, P., Jiménez, J.A. and Justice, D. 1999. Population dynamics of mangrove *Avicennia bicolor* on the Pacific coast of Costa Rica. Wetland Ecology and Management 7: 113-120.

Duke, N.C. 2001. Gap creation and regenerative processes driving diversity and structure of mangrove ecosystems. Wetlands Ecology and Management 9: 257-269.

Ellison, A.M and Farnsworth, E.J. 1997. Simulated sea level change alters anatomy, physiology, growth, and reproduction of red mangrove (*Rhizophora mangle* L.). Oecologia 112: 435-446.

Ellison, J.C. 1993. Mangrove retreat with rising sea level, Bermuda. Estuarine, Coastal and Shelf Science 37: 75–87

Ellison, J.C. 1999. Impacts of sediment burial on mangroves. Marine Pollution Bulletin 37: 420-426.

Elster, C. 2000. Reason for reforestation success and failure with the mangrove species in Colombia. Forest Ecology and Management 131: 201-214.

Elster, C., Perdomo, L. and Schnetter, M.L. 1999. Impact of ecological factors on the regeneration of mangroves in the Ciénaga Grande de Santa Marta, Columbia. Hydrobiologia 413: 35-46.

Flos, S.J. 1993. Estimations of freshwater flow through the Pak Phanang River. M.Sc. Thesis, University of Humberside, England.

Furukawa, K. and Baba, S. 1997. Effect of sea level rise on Asian mangrove forests. In: Proceedings of the APN/SURVAS/LOICZ Joint Conference on Coastal Impacts of Climate Change and Adaption in the Asia-Pacific Region. Kobe, Japan - November 14-16, 2000.

Havanond, S. and Maxwell, G.S. 1993. Destruction of Rhizophora seedlings by Hermit crabs in a mangrove forest plantation in Thailand. The Eight National Seminar on Mangrove Ecology and Management. Surat Thani, Thailand, August Volume I part II pp20-24.

High Performance Systems (HPS), 1997. STELLA technical documentation. High Performance Systems, Hanover, New Hampshire.

Hull, C.H.J and Titus, J.G. (editors) 1986. Greenhouse effect, sea level rise, and salinity in the Delaware Estuary. United States Environmental Protection Agency Report.

Hutchings, P. and Saenger, P. 1987. Ecology of Mangroves. University of Queensland Press, St Lucia.

Intergovernmental Panel on Climate Chance (IPCC). 1998. The Regional Impacts of Climate Change: An Assessment of Vulnerability. IPCC Special Report.

ISME/GLOMIS. River damming and changes in mangrove distribution. ISME/GLOMIS Electronic Journal. Volume 2, July 2002.

Japan International Corporation Agency (JICA). 1987. Report on the basic design for constructing the Nakhon Si Thammarat fishing port in the Kingdom of Thailand. Japan International Corporation Agency, 259 pp.

Jirawattanapun, S., Patanaponpaiboon, P., Aksornkoae, S. and Pliyavuth, C. 2002. Effects of salinity on distribution of *Sonneratia caseolaris* and *Soneratia alba*. In: Proceeding of the 12th Thailand National Seminar on Mangrove Ecology, 28-30 August 2002, Nakhon Si Thammarat, Thailand, National Research Council of Thailand, III3: 1-8 (in Thai).

Kairo, J.G., Dadough-Guebas, F., Gwada, P.O., Ochieng, C. and Koedom, N. 2002. Regeneration status of mangrove forests in Mida Creek, Kenya: a compromised or secured future. Ambio 31: 562: 568.

Kaly, U.L. and Jones, G.P. 1998. Mangrove restoration: a potential tool for coastal management in tropical developing countries. Ambio 27:656-661.

Lee, S.K., Tan, W.H. and Havanond, S. 1996. Regeneration and colonisation of mangrove on clay-filled reclaimed land in Singapore. Hydrobiologia 319: 23-35.

Mazda, Y., Magi, M., Kogo, M. and Hong, P.N. 1997. Mangroves as a coastal protection from waves in the Tong King delta, Vietnam. Mangroves and Salt Marshes 1:127-135.

McKee, K.L. 1995. Seedling recruitment patterns in a Belizean mangrove forest: effects of establishment ability and physico-chemical factors. Oecologia 101: 448-460.

McMillan, C. 1971. Environmental factors affecting seedling establishment of the Black mangrove on the Central Texas coast. Ecology, 52: 927-930.

Meteorological Department 2004. Rainfall data 1971-2000. The Meteorological Department, Bangkok, Thailand.

Panapitukkul, N., Duarte, C.M., Thampanya, U., Terrados, J., Keowvongsri, P., Geertz-Hansen, O., Srichai, N. and Boromthanarat, S. 1998. Mangrove colonization: Mangrove progression over the growing Pak Phanang (SE Thailand) mud flat. Estuarine, Coastal and Shelf Science 47: 51-56.

Patanaponpaiboon, P. and Pliyavuth, C. 2001. Effects of water salinity on growth of some mangrove seedlings. In: Proceeding of the 11th Thailand National Seminar on Mangrove Ecology, 9-12 July 2000, Trang, Thailand, National Research Council of Thailand, I5: 1-11 (in Thai).

Patterson, S., McKee, K.L. and Mendelssohn, I.A. 1997. Effect of tidal inundation and *Avicennia germinans* seedling establishment and survival in a sub-tropical mangal/salt marsh community. Mangroves and Salt Marsh 1: 103-111.

Purba, M. 1991. Impact of high sedimentation rate on the coastal resources of Segara Anakan, Indonesia, p.143-152. In L.M. Chou, T.E. Chua, H.W. Khoo, P.E. Lim, J.N. Paw, G.T. Silvertre, M.J. Valencia, A.T. White and P.K. Wong (eds) Towards an integrated management of tropical coastal resources. ICLARM conference proceedings 22, 445 p.

Putz, F.E. and Chan, H.T. 1986. Tree growth, dynamics, and productivity in a mature mangrove forest in Malaysia. Forest Ecology and Mangement 17: 211-230.

Rabinowitz, D. 1978. Dispersal properties of mangrove propagules. Biotropica 10: 47-57.

Saito, Y. 2001. Deltas in Southeast and East Asia: Their evolution and current problems. In: Mimura, N. and Yokoki, H. (ed.), Global Change and Asia Pacific Coasts. Proceeding of APN/SURVAS/LOICZ Joint Conference on Coastal Impacts of Climate change and Adaptation in the Asia-Pacific Region, APN, Kobe, Japan, November 14-16, 2000, pp.185-191.

Sherman, R.E, Fahey, T.J, and Battles, J.J. 2000. Small-scale disturbance and regeneration dynamics in a neotropical mangrove forest. Journal of Ecology 88: 165-178.

Smith, T.J. III, Chan, H.T., McIvor, C.C. and Robblee, M.B. 1989. Comparison of seed predation in tropical tidal forests from three continents. Ecology 70: 146-151.

Terrados, J., Thampanya, U., Srichai, N., Keowvongsri, P., Geertz-Hansen, O., Boromthanarat, S., Panapitukkul, N. and Duarte, C.M., 1997. The effect of increased sediment accretion on the survival and growth of *Rhizophora apiculata* seedlings. Estuarine, Coastal and Shelf Science 45: 697-701.

Thampanya, U., Vermaat, J.E. and Duarte, C.M. 2002a. Colonization success of common Thai mangrove species as a function of shelter from water movement. Marine Ecology Progress Series 237: 111-120.

Thampanya U., Vermaat, J.E. and Terrados, J. 2002b. The effect of increasing sediment accretion on the seedlings of three common Thai mangrove species. Aquatic Botany 74: 315-325.

Thom, B.G. 1984. Coastal landforms and geomophic processes: 3-17. In: The mangrove ecosystem research method. (Snedaker S.C. and Snedaker, J.G., eds). Unesco, Paris.

Verheyden, A., Kairo, J.G., Beeckman, H. and Koedam, N. 2004. Growth ring, growth ring formation and age determination in the mangrove *Rhizophora mucronata*. Annals of Botany 94: 59-66.

Wechakit, D. 1987. Growth and yield of *Rhizophora apiculata* planted in private forest, Samut Songkram province, Thailand. M.Sc. Thesis. Kasetsart University, 71pp.

Woodroffe, C.D. 1992. Mangrove sediments and geomorphology. In: D. Alongi and A. Robertson (ed.), Tropical Mangrove Ecosystem. American Geophysical Union, Coastal and Estuarine Studies, p7-41.

World Wildlife Fund (WWF) website. 2005. Indochina mangroves. Last access September 4, 2005.

Yulianto, E., Sukapti, W.S., Rahardjo, A.T., Noerdi, D., Siregar, D.A., Suparan, P. and Hirakawa K. 2004. Mangrove shoreline responses to Holocene environmental change, Makassar Strait, Indonesia. Review of Palaeobotany 131: 251-268.

Chapter 7

Conclusion and general discussion

Introduction

Mangroves grow between land and sea where riverine sediment deposits and where exposure to water turbulence is common. These two factors, therefore, have a great impact on coastal dynamics and mangroves. This dissertation has attempted to quantify this experimentally and then integrate their relative importance. Our focus was on the early stages of the mangrove life cycle since we assumed these to be most sensitive. As it had its roots in the Coastal Ecosystems Response to Deforestation-derived Siltation in Southeast Asia project (CERDS, e.g. Duarte et al. 1998; Panapitukkul et al. 1998; Terrados et al. 1998; Kamp-Nielsen et al. 2002), this work started off with an emphasis on sedimentation and seedling burial (Chapter 3, based a.o. on Terrados et al. 1997), but it expanded to include coastal hydrodynamics (Chapter 2) and added the larger spatial scale of the whole southern Thai coastlines to our original focus on mangrove-dominated bays such as Pak Phanang. This is justified as it relates the relevance of our findings in bays to the whole coast. Chapter 5 provided this larger scale perspective. It was originally planned to develop demographic models that could be useful for coastal management. Despite the variable quality of the data available, three simple models were developed and satisfyingly validated in chapter 6. Alongside the collection of data for chapter 5 and 6, it appeared necessary to compile a data set on mangrove growth equations. These have been brought together in chapter 4.

This concluding chapter summarizes the main findings from chapter 2-5 where different data sets have been collected and analyzed separately. Since the modelling chapter 6 has brought together all these data and analyzed them in concert, its main conclusions are then discussed particularly at length. This chapter concludes with practical recommendations for mangrove forestry and coastal zone management in SE Asia.

Turbulence

The impact of exposure to water turbulence on mangrove seedlings was quantified in an *in situ* experiment (Chapter 2). The results reveal that water turbulence, indeed, has impacts on growth and survival of the three mangrove species studied: *Avicennia alba*, *Rhizophora mucronata* and *Sonneratia caseolaris**. During the one year experiment, mortality of *Sonneratia* increased with increasing water movement in the monsoon period while mortality of *Avicennia* was spread equally. *Rhizophora* survived poorly in the most exposed plot. Its causes of mortality are mostly from uprooting by strong currents or being broken by water-borne objects. Considering seedling growth, the first two species exhibited a higher height increment in more exposed plots. Likewise, internode production of all three species was also highest in the most exposed plots. This is probably a rapid growth response to cope with the harsh environment. *Avicennia* and *Sonneratia*, the two pioneer species, were found to be able to grow and survive well in a high exposure environment. These findings are consistent with the role of the three species in the successional series, suggesting exposure to water movement to be an important determinant of mangrove colonization and succession.

Sediment deposition

It can be said with confidence from our burial experiment of the three mangrove species (Chapter 3) that variation in sediment accretion has affected growth and survival of mangrove seedlings established on the mudflats. In this experiment, the species used were the same as the previous chapter, except that *Avicennia officinalis* was used instead of *Avicennia alba* due to the availability of seedlings existing in the study area. The numbers of surviving seedlings of *Avicennia* and *Sonneratia* were highly affected by burial and their survival decreased with increasing sediment accretion. The seedlings receiving 32 cm of sediment had the highest mortality (100% in *Avicennia*, 70% in *Rhizophora* and 40% in *Sonneratia*). Among the three taxa, sediment burial had significant effects on seedling growth in terms of height in *Avicennia* and *Sonneratia* and height increment was lowest in the highest experimental sediment accretion (32 cm). Annual internode production, however, was not significantly affected by burial in any species. *Avicennia* was most sensitive to sediment burial, especially, with sediment level higher than 16 cm. The results also illustrate that a fast growing species, in this case *Sonneratia*, may well be able to survive and colonize or be planted in areas where abrupt high sedimentation is possible. Therefore, reforestation programs with mainly ecological purposes could favourably use this species more often or use as a pioneer species prior to the plantation one of a higher economic value such as *Rhizophora*. Besides the contribution of high primary production from faster leaf decomposition rate (Ashton et al. 1999), *Sonneratia* wood has medium quality for firewood and charcoal (Taylor et al. 2003). Furthermore, its young flowers and fruits are edible and are sold in local market (Thampanya pers. observ.; Plathong & Sitthirach 1997; FAO 2005).

Mangrove tree growth equations

We compiled regression equations of Height-trunk diameter at breast height (DBH)-age of the three mangrove species (Chapter 4) to be used as a tool for an estimation of mangrove age from height or DBH measurement. Overall, height-age and DBH-age data were fitted satisfactorily with linear regression models and the two regressions were found to describe the observed data well. The resulting regression equations were used for the quantification of mangrove progression in the subsequent chapter. This compilation of equations includes a considerable amount of grey literature. It can, therefore, be useful to forest managers in SE Asia, and probably beyond. For example, it was found that fitted DBH-age and height-age curves were identical across 800 km for *Rhizophora apiculata*, suggesting the potential for extrapolation and wider-scale applicability of these equations. Furthermore, the differences in elongation and wood accumulation confirmed well with the ecological status of the taxa as early colonizers or preferred forestry target species (Figure 4.3).

Coastal change

Figure 7.1. Coastal erosion at Ban Chai Thale, Pak Phanang District, Nakorn Si Thammarat Province, 4 December 2004 (photo: U. Thampanya).

Chapter 5 examined coastal dynamics along Thailand's coastlines both on the Andaman Sea coast and the Gulf of Thailand coast. The relationship between mangrove presence and changes in coastal area was assessed. Available quantitative data on changes of coastal segments in southern Thailand (erosion and accretion) as well as possible factors responsible for these changes were compiled. It was found that, the coastlines of Thailand have been experiencing both erosion and accretion. The more developed (i.e. dense aquaculture ponds and settlements) Gulf coast to the east was found to be most dynamic: it had higher rates of progression in sheltered, mangrove-dominated coastal segments and higher erosion in exposed segments (Figure 7.1). Overall, Southern Thailand has been loosing its coastal area by a rate of 1.3 m y^{-1} during the period of 1967-1998. The total area losses amounted to 0.91 and 0.25 km^2 y^{-1} for the east and the west coast, respectively (Figure 5.7). Mangrove-dominated coastlines exhibited less erosion than non-vegetated ones. Mean erosion rate of coastal segments with mangroves was lower than those without mangroves (2.3 m y^{-1} vs 3.3 y^{-1}, Figure 5.6). Possible factors underlying coastal changes were examined. It was found that erosion increased with increased area of shrimp farms in the coastal lowland, increased fetch to the prevailing monsoon across the sea, and when dams reduced riverine inputs from hinterland. This study underpins that, next to sedimentation and exposure to turbulence at a bay-scale, anthropogenic activities and economic development in the coastal area and beyond, at the catchment scale, are important factors underlying coastal change. Less sediment delivery to the coastal area accelerates coastal erosion. Finally, the presence of mangroves may not prevent coastal erosion in the long run but will probably help in reducing present rates of erosion.

Demographic models

To obtain an overall understanding of the mangrove colonization process, three demographic models were developed, one for *Avicennia*, *Rhizophora* and *Sonneratia* each (Chapter 6). A model is composed of five stages of the mangrove life cycle and includes the effects of key environmental factors: water turbulence, herbivory, sedimentation, salinity and drought condition (Figure 6.6 and 6.7). The simulation of the three species models produced reasonable to good estimates of colonized areas. The area colonized after 30 years was comparable to the real historical situation of the mangrove forests in Pak Phanang Bay (the main study area of this thesis). Our validation with the field observations showed a significant difference only for *Rhizophora*. Sensitivity testing suggested that the most sensitive parameters were seedling herbivory notably by crabs and water turbulence. The model runs demonstrated the high success in colonization on the mudflat of *Avicennia* species. *Sonneratia* was second in simulated area colonized while the area covered by *Rhizophora* was limited. This work has shown that apart from the differences in responsive capacity to the environmental factors, biological traits, such as the number of seeds produced, propagule buoyancy and preference by herbivory, contributed greatly to variation in magnitude of colonization success of the different species. This confirms the observed sensitivity by herbivory, where sensitivity of *Sonneratia* was approximately four-fold higher than *Avicennia* and 20-fold higher than *Rhizophora* (Figure 6.10).

In an extrapolation exercise to predict the impact of global climate change and coastal development on mangrove dynamics, a range of scenarios was applied. The model simulations illustrated that accelerated sea level rise will reduce the success in colonization of all three mangrove taxa and particulary of *Avicennia* and *Sonneratia* (Table 7.1; Figure 6.11). In addition, 50% decrease in rainfall, will decrease colonization of the three species since less exportation of propagules to the mudflat occurs and seedling desiccation is prevalent. Freshwater cut-off by full river damming will adversely affect *Sonneratia*: its colonization will almost totally cease. However, this situation was found to be favorable for *Avicennia* as its colonization increases dramatically. Our model demonstration reveals that change in environmental factors may have both negative and positive impacts on particular species. Since the sea level rise scenario applied has a high probability to materialize in reality, we can predict declining colonization success rates and hence reduced mangrove cover in the longer run with some confidence; that is, unless new areas for colonization become available further inland.

Table 7.1. Predicted colonization success (%; relative to present baseline) of three mangrove taxa under changes in environmental conditions. Presented are four selected scenarios applied in Chapter 6.

species	SLR	damming	double rainfall	50% lower rainfall
Avicennia	60	140	20	60
Rhizophora	90	60	150	50
Sonneratia	60	5	20	30

All in all, this study has demonstrated that changes in sedimentation, variation in hydrodynamics and biotic factors, such as seed and seedling herbivory together affect the colonization rates of mangroves along the coastlines of SE Asia. Local rates of sedimentation may adversely affect survival, similar to high exposure to turbulence. The larger scale reduction in sediment delivery to the coast results in serious coastal erosion and declines in available habitat. Our integrative modelling suggests that colonization success was notably sensitive to turbulence and seedling herbivory. Scenario runs suggests that anticipated SLR will negatively affect mangrove colonization, as will increased damming across inflowing rivers. The latter may be positive only for the most salt-tolerant taxon *Avicennia*.

Recommendations

The main objective of this thesis was to gain insight into the impact of sedimentation and related factors on mangrove colonization success and coastal dynamics. The thesis resulted in an integrated perspective that can be used as management guidelines and decision-making for successful and efficient coastal zone management. This section presents some recommendations for coastal management practice and further research in the field of mangrove ecology and coastal dynamics.

Management practice

It is evident from chapter 5 that Thailand has been loosing substantial tracts of its coastal land over the past decennia. This situation is likely to intensify due to global climate change in the next decades. Therefore, its consequences and possible remedial measures should urgently be taken into account in coastal management schemes.

Detailed assessment and mapping of vulnerable areas to extensive erosion should be carried out at the earliest. Adaptation literature (e.g. Nicholls & Klein 2005) suggests that managed retreat could be one of the more viable strategies. Consideration of range possible management strategies for specific coastal segments would require a careful balancing of local coastline and sediment dynamics. A possible, inexpensive and natural-friendly strategy is mangrove plantation. Although mangroves may not prevent coastal erosion ultimately, the existence of a 300-500 m belt of mangroves may help reducing erosion rate (Winterwerp et al. 2005). This can be done by planting the seedlings of the species present in adjacent areas, possibly initially - with sheltering support from e.g. stakes, rock mounds or bricks (Figure 7.2 and 7.3). In some cases filling with fluvial soils may be needed. Introduction and maintenance of mangrove plantations in vulnerable areas is probably beneficial for long-term coastal protection to gradual, continuous erosion or also more severe events such as a tsunami.

Similarly, a possible strategy to sustain the coastal zone area is to enhance the total area covered by mangroves. The easiest and least expensive way to achieve this goal is to assist natural mangrove colonization in sheltered coastal segments by providing or enhancing seedling fluxes to the area, protecting seedlings from herbivory and increasing propagule retention time with artificial shelters. In addition, mangrove afforestation on unoccupied mudflats should emphasize an ecological perspective and use the suitable and fast growing species such as *Avicennia* and *Sonneratia* in stead of a commonly-used *Rhizophora* or plant these two species at the beginning to prepare shelters for the late successional species.

Figure 7.2. Existing mangrove stands associated with rock mounds helps in reducing erosion of   a coastal segment on the east coast of peninsular Thailand, 15 May 1996 (photo: CORIN).

Figure 7.3. Mangrove seedlings were planted and fixed with wooden stakes at Songkhla lake outlet, Southern Thailand (photo: U. Thampanya).

More importantly, government should issue and enforce legislation to control aquaculture industry in the coastal zone. For instance, the cultivation ponds must be located at the certain minimum distance from the coastline (Figure 7.4a and 7.4b) or located at least behind 300-500 mangrove or beach forests and some mitigation measures for coastal erosion must be prepared. Such legislation would profitably be accompanied by monitoring and - should be enforced by authorized government agencies such as the Department of Fisheries and the Ministry of Natural Resources.

Figure 7.4. (a) and (b) Present of shrimp ponds located only few distance along the coast of Nakron Si Thammarat Province, Southern Thailand (photo: CORIN).

Gaps in knowledge

(1) Mangrove ecology

Notably from chapter 6 it became apparent that several aspects of mangrove ecology are poorly covered as yet. These are: the quantification of mangrove reproduction, propagule transportation, retention time and establishment rate in relation to local hydrodynamics and quantification of seed germination and seedling herbivory *in situ*. Furthermore, our modelling of early colonization filled a gap compared to the previous published models (Chen & Twilley 1998; Berger & Hildenbrandt 2000; Berger et al. 2002). Bringing these different approaches together would be extremely useful as it will be an exercise in resolving scaling problems as well as matching conceptual models.

Additionally, the standard error of mangrove age estimation (Chapter 4) was found to be high, particularly, for *Sonneratia*. This was because data availability was quite limited for this species on actual mangrove life span as well as the known age or planting date of the old trees. Also, growth data varied with location and environment, although the similarity among widely separate stands of *Rhizophora apiculata* appears promisingly useful. More research on mangrove growth-age relationship would be beneficial for mangrove forestry and management. Our data mining suggests that more such data collections may well be present in the region that – could usefully be made publicly available.

(2) Sedimentation and sea level rise

In this study, data on the sediment accretion of the Pak Phanang mangrove forest were measured during 1998-1999 using 30 artificial measuring poles. An average accretion rate of 2.0 ± 0.2 cm y^{-1} was observed, which must be interpreted as an instantaneous, gross rate that does not incorporate compaction or subsidence. This rate compares well to the estimated sea level rise of 1-2 cm y^{-1} (chapter 6), as well as to accretion in salt marshes (e.g. Day et al. 1999). A proper longer-term assessment would also involve a consideration of Holocene transgression, carbon-dated sedimentation history as well as peat formation (e.g. Ellison, 1993).

(3) Integrated management

Since coastal ecosystems are complex and there is a linkage between coastal and terrestrial ecosystems, understanding of these inter-related ecosystems and the impacts of human interventions would be beneficial for efficient management planning and practices. Particularly in coastal zone management, social-ecological resilience (Adger et al. 2005) and on-going capacity building on an integrated perspective to prevent sectoral uni-dimensional solutions should become more common-place in local and regional governance systems. An example is illustrated in Pak Phanang Bay where river damming for irrigation purposes (Figure 7.5) has impeded seasonal migration of fish and crustacean species causing an economically value brackish water fisheries to disappear from the area (Sirimantaporn & Boontae 2001). Catchment-scale research which includes natural sciences, economics and social sciences (LOICZ 2005; Vermaat et al. 2005), therefore, would offer the necessary answer to questions at this scale.

Figure 7.5. An irrigation dam has been constructed at approximately 9 km from the Pak Phanang river mouth and started its operation since October 2000 (phot: Royal Irrigation Dept).

References

Adger W.N., Hughes, T.P., Folke, C., Carpenter, S.R. and Rockström, J. 2005. Social-ecological resilience to coastal disasters. Science, 309: 1036-1039.

Ashton, E.C, Hograth, P.J. and Ormond, R. 1999. Breakdown of mangrove leaf little in managed mangrove forest in Peninsular Malaysia. Hydrobiologia, 413: 77-88.

Berger, U. and Hildenbrandt, H. 2000. A new approach to spatially explicit modelling of forest dynamics: spacing, aging and neighbourhood competition of mangrove trees. Ecological Modelling 132: 287-302.

Berger, U., Hildenbrandt, H. and Grimm, V. 2002. Towards a standard for the individual-based modeling of plant populations: self-thinning and the field-of-neighborhood approach. Natural Resource Modeling 15: 39-54.

Chen, R. and Twilley, R.R. 1998. A gap dynamic model of mangrove forest development along gradients of soil salinity and nutrient resources. Journal of Ecology 86: 37-51.

Day, Jr J.W., Rybczyk, J., Scarton, F., Rismondo A., Are, D. and Cecconi, G. 1999. Soil accretionary dynamics, sea level rise and the survival of wetlands in Venice Lagoon: a filed and modeling approach. Estuarine, Coastal and Shelf Science 49: 607-628.

Duarte, C.M., Geertz-Hansen, O., Thampanya, U., Terrados, J., Fortes, M.D., Kmap-Neilsen, L., Borum, J. and Boromthanarat, S. 1998. Relationship between sediment conditions and mangrove Rhizophora apiculata seedling growth and nutrient status. Marine Ecology Progress Series 175: 277-283.

Ellison, J.C. 1993. Mangrove retreat with rising sea level, Bermuda. Estuarine, Coastal and Shelf Science 37: 75–87

FAO, 2005. Mangroves: what are they worth? FAO Corporate Document Repository, Forestry Department. FAO website, URL: www.fao.org

Kamp-Nielsen L.,Vermaat J.E.,Wesseling I.,Borum J. and Geertz-Hansen O. 2002. Sediment properties along gradients of siltation in South-east Asia. Estuarine, Coastal and Shelf Science 54 : 127-137.

Land-Ocean Interactions in the Coastal Zone (LOICZ). 2005. Science plan and implementation strategy. IGBP Report 51/ IHDP Report 18. 60pp.

Nicholls, R.J. and Klein R.J.T., 2005. Climate change and coastal management on Europe's coast. In: Vermaat, J., Bouwer, L., Turner, K. and Salomons, W., (Eds). 2005. Managing European coasts: past, present and future. Springer-Verlag, Berlin Heidelberg, pp 199-226.

Panapitukkul, N., Duarte, C.M., Thampanya, U., Terrados, J., Keowvongsri, P., Geertz-Hansen, O., Srichai, N. and Boromthanarat, S. 1998. Mangrove colonization: Mangrove progression over the growing Pak Phanang (SE Thailand) mud flat. Estuarine, Coastal and Shelf Science 47: 51-61.

Plathong, J. and Sitthirach, N. 1997. Traditional and current uses of mangrove forests in southern Thailand. Wetlands International-Thailand Programme/PSU, Publication No. 3.

Sirimantaporn, P. and Boontae, J. 2001. Survey of bio-fishery and aquatic ecosystem in Pak Phanang river basin (before and after the construction of Pak Phanang dam). Thailand National Institute of Coastal Aquaculture, Technical Report, 10pp.

Taylor, M., Ravilious, C. and Green, E.P. 2003. Mangroves of East Africa, UNEP-WCMC Report, 26pp.

Terrados, J., Thampanya, U., Srichai, N., Keowvongsri, P., Geertz-Hansen, O., Boromthanarat, S., Panapitukkul, N. and Duarte, C.M. 1997. The effect of increased sediment accretion on the survival and growth of *Rhizophora apiculata* seedlings. Estuarine, Coastal and Shelf Science 45: 697-701.

Terrados, J., Duarte, C.M., Fortes, M.D., Borum, J., Agawin, N.S.R., Bach, S., Thampanya, U. and Kmap-Neilsen, L. 1998. Changes in community structure and biomass of seagrass communities along gradients of siltation in SE Asia. Estuarine, Coastal and Shelf Science 46: 757-768.

Vermaat, J., Bouwer, L., Turner, K. and Solomons, W., (Eds). 2005. Managing European coasts: past, present and future. Springer-Verlag, Berlin Heidelberg, 387pp.

Winterwerp, J.C., Borst, W.G. and de Vries, M.B. 2005. Pilot study on the erosion and rehabilitation of a mangrove mud coast. Journal of Coastal Research 21: 223-230.

Samenvatting

Mangroves en sediment dynamiek aan de kusten van Zuid Thailand

Mangrovebossen komen voor langs tropische en subtropische kusten, met name in beschutte delta's en baaien die zoet water en slib van rivieren ontvangen. Mangrovebossen staan bekend om de belangrijke ecologische en sociaal-economische functies die ze voor de kustbewonende bevolking vervullen. Zo fungeren ze als natuurlijke paaiplaats en leefgebied voor vele soorten vis en schaaldieren. Het hout wordt uitgebreid gebruikt als constructiemateriaal, als brandhout en voor het maken van houtskool. De wortelsystemen bevorderen opslibbing en de afzetting van slikplaten, en stabiliseren daarmee de kustlijn. Zo leveren mangrovebossen ook beschutting bij jaarlijks voorkomende moessonstormen en cyclonen, en zelfs gedurende tsunamis, zoals in 2004 in Zuid Thailand is gebleken.

De huidige toestand van deze waardevolle natuurlijke hulpbron is echter kritiek, met name in Zuidoost Azië, waar het oppervlakte van mangrovebossen naar schatting is gehalveerd in de afgelopen 35 jaar. De belangrijkste oorzaken voor deze achteruitgang zijn de geleidelijke uitbreiding van kustnederzettingen door populatiegroei, grootschalige omvorming tot vijvers voor garnalen- en visteelt, kusterosie, gebrek aan bewustzijn en gebrekkige handhaving van wetten en regels. Satellietbeelden van de Thaise kust laten overduidelijk zien hoe wijdverbreid de aquacultuur is. Inmiddels zijn zowel regering als het publiek zich veel bewuster van het belang van mangroves, hetgeen tot een reeks herbebossingsprojecten heeft geleid. Herbebossing bleek vaak echter verre van eenvoudig en werd ook vertraagd door landrecht-disputen. Daarom is geregeld voor herbeplanting uitgeweken naar nieuw opgeslibde slikplaten. Hier bleken de nieuwe plantages echter vaak niet of weinig succesvol, waarschijnlijk omdat niet was nagegaan of de gebruikte soorten geschikt waren voor betreffende standplaats en omdat de mogelijkheid van natuurlijke rekolonisatie niet van te voren was onderzocht.

Ten bate van een verantwoord mangrove- en kustbeheer richtte dit proefschrift zich op een verbetering van het begrip van de factoren die mangrove-kolonisatie beïnvloeden, in samenhang met de veranderingen die langs tropische mangrovekusten optreden. In de eerste plaats is experimenteel in het veld onderzocht wat het effect van sedimentatie en waterbeweging is op de overlevingskansen van vivipare kiemplanten van drie algemene mangrove-taxa, *Avicennia, Rhizophora* en *Sonneratia*. Vervolgens is het perspectief verbreed naar de kustdynamiek in geheel Zuid Thailand gedurende de laatste 35 jaar. Tenslotte is een en ander samengebracht in eenvoudige demografische modellen die het kolonisatieproces van mangrove soorten simuleren.

Twee veldexperimenten toonden aan dat mangrove-kiemplanten ook na succesvolle vestiging nog een aanzienlijke kans lopen om door plotselinge opslibbing of intensieve golfslag af te sterven. Sterfte nam toe bij toenemende opslibbing: *Avicennia* kiemplanten overleefden begraving met 32 cm sediment niet, terwijl van *Rhizophora* en *Sonneratia* nog 30% en 60% deze behandeling overleefden. De laatste soort was het minst gevoelig voor begraving en vertoonde een snelle lengtegroei. Blootstelling aan hevige golfwerking bleek nadeling voor de overleving van *Rhizophora* kiemplanten, maar de andere twee soorten deden het onder dergelijke omstandigheden juist beter. Dit komt overeen met de status van deze soorten als vroege pioniers en suggereert dat ze het best geschikt zouden zijn voor herkolonisatie- projecten, waarbij *Sonneratia* dan de voorkeur zou hebben op plaatsen waar plotselinge en/of aanzienlijke opslibbing mogelijk is.

De lange termijn veranderingen langs de Zuid Thaise kust bleken sterker te zijn geweest aan de oostkant, langs de Thaise Golf, dan langs de westkust. Erosie vond plaats langs 29% van de oostkust maar slechts langs 11% van de westkust en heeft geleid tot gemiddelde landverliezen van respectievelijk 91 en 25 ha per jaar. De belangrijkste oorzaken voor dit verlies aan land waren verlies aan mangrove-oppervlakte, toename in vijvers voor garnalenteelt, afgenomen slibtransport van land naar zee door een toename aan dammen in rivieren, en blootstelling aan sterke moessonwinden. Waar mangroves de kust domineerden was minder erosie .Landaanwas vond slechts plaats langs beschutte, met mangrove bedekte kust segmenten. Op dergelijke plaatsen groeide de kust met 37 ha per jaar in het oosten en met 5 ha in het westen. Alles bij elkaar genomen is er sprake van een netto landverlies in Zuid Thailand van 74 ha per jaar gedurende de laatste drie decennia. Voldoende sedimentaanvoer en een goed ontwikkelde mangrovekust blijken aanzienlijk bij te dragen aan de stabiliteit van de kustlijn.

De demografische simulatiemodellen lieten zien dat met name herbivorie (kiemplanten zijn kwetsbaar voor krabben) en waterbeweging factoren zijn die een rol spelen bij succesvolle kolonisatie in beschutte baaien, een typische mangrove habitat; geleidelijke opslibbing en saliniteit hadden minder invloed. *Avicennia* bleek het makkelijkst open slibvlaktes te koloniseren, gevolgd door *Sonneratia*, terwijl *Rhizophora* daarin minder succesvol was. Veranderende milieu-omstandigheden zijn gesimuleerd met behulp van een serie plausibele scenarios. Zo leidde versnelde zeespiegelstijging tot een geringer succes voor alle drie de soorten, en verminderde zoetwateraanvoer (meer dammen) verhoogde de kansen voor *Avicennia* maar verminderde deze juist voor *Sonneratia*.

Grosso modo laat deze studie zien dat herkolonisatie-succes van mangroves met name door ecologische en hydrologische factoren bepaald wordt. Kiemplantmortaliteit door herbivorie en waterbeweging blijken in grote mate de overlevingskansen van jonge mangroveboompjes te bepalen. De drie onderzochte taxa vertoonden aanmerkelijke verschillen in kolonisatie-capaciteit. Geleidelijke sedimentatie is niet negatief voor individuele kiemplanten en een positieve sedimentatiebalans maakt vooral in beschutte baaien nieuwe oppervlakten slik beschikbaar voor kolonisatie. Afgenomen slibtransport door rivieren naar zee zal erosie in kritische gebieden waarschijnlijk verhevigen. Voor een duurzaam kustbeheer is meer aandacht nodig voor de zorgvuldige afweging van ecologische en sociaal-economische belangen.

Resumen

Dinámica de Sedimentos y Ecosistemas de Manglar en las Costas del SUr de Tailandia

Los bosques de manglar son un tipo específico de bosques perenifolios que crecen en la costa de las regions tropicales y subtropicales, particularmente en deltas y bahías que reciben descargas fluviales de agua ducle y sedimentos. Los bosques de manglar cumplen importantes funciones en el ecosistema que aportan valores socio-económicos a las comunidades costeras. Por ejemplo, son areas naturals para la puesta y cría de muchas especies de peces y crustáceos. Su madera es utilizada por las comunidades locales como material de construcción, combustible y para hacer carbón, y sus sistemas radicals estabilizan los sedimentos. Así, estos bosques ejercen importantes funciones de protección del litoral durante tormentas, cyclones y tsunamis. Esta función se reflejó de forma dramatica durante el tsunami de 2004, en el que la presencia de bosques de manglar mitigó la devastación causada, en terminus de vidas humanas y propiedades, en las comunidades que se situaban tras los manglares.

Estos valiosos recursos se encuentran en un estado crítico, particularmente en el SE Asiático, donde la extension de los bosques de manglar ha disminuído a menos de la mitad en 35 años. Las causas más importantes de esta pérdida son la expansion de la población, la conversion a estanques de acuicultura, la erosion costera, y la falta de concienciación sobre su importancia junto con estratégicas de conservación isnfucientemente desarrolladas. En Tailandia, por ejemplo, imagines de satellite recientes apuntan a un gran desarrollo de estanques de acuicultura a lo largo de la costa del Golfo de Tailandia debido a la ausencia de restricciones normativas. Más recientemente, el aumento de la concienciación del gobierno y del público en cuanto a la importancia de los manglares se ha traducido en un aumento de los esfuerzos de restauración y reforestación de manglares. La restauración de areas de manglares degradadas se ha encontrado con un número de problemas, incluidos conflictos en torno a la propiedad de las tierras. Por ello, muchos proyectos se han redirigido hacia la reforestación de nuevas llanuras intermareales. Estos proyectos han teniddo un éxito variable y a veces han fracasado en sus esfuerzos, probablemente porque los proyectos se abordaron sin considerar previamente qué especies eran más apropiadas, las técnicas a emplear, y las capacidades para la recolonización natural de estas comunidades.

A fin de generar información útil para la gestión de los manglares y la zona costera, esta disertación tiene como objetivo generar una major comprensión de los factors que afectan el éxito de la colonización y aquellos que inciden sobre los cambios en la zona costera. En primer ligar, se evaluó experimentalmente a escala de la bahía el impacto de la sedimentación y turbulencia sobre la supervivencia de las plántulas y el crecimiento de 3 géneros de manglares dominantes, *Avicennia*, *Rhizophora* y *Sonneratia*. Esta inviestigación se amplió para evaluar la dinámica costera en el Sur de Tailandia, a través de una síntesis de información en las zones de estudio para el periodo 1961-2000. Finalmente, se desarrollaron modelos sencillos de colonización para similar el éxito del proceso de colonización de manglares.

Los dos experimentos mostraron que tras el establecimiento en los sedimentos intermareales, las plántulas podían morir debido a la turbulencia del agua y a su enterramiento por eventos de deposición de sedimentos. La mortalidad aumentó al aumentar las tasas de aporte de sedimentos y ninguna de las plántulas de Avicennia sobrevivió bajo la tasa de enterramiento más alta de 32 cm, mientras que las de *Rhizophora* y *Sonneratia* mostraron una cierta supervivencia a esas tasas de enterramiento (respectivamente 30% y 60%). De entre estos tres géneros, *Sonneratia* fue la menos afectada por el enterramiento y mostró una rápida tasa de crecimiento. Las plántulas de *Rhizophora* mostraron la menor supervivencia en localidades altamente expuestas (baja densidad de plantas) comparada con localidades más protegidas. Estos resultados confirman el comportamiento de *Avicennia* y *Sonneratia* como especies pioneras que puden colonizar llanuras intermareales, seguidas de *Rhizophora* en la sucesión. Sin embargo, *Sonneratia* estaría major adaptada a areas sujetas a movimientos bruscos de sedimentos que *Avicennia*.

La evaluación de cambios en la zona costera del Sur de Tailandia mostró que la costa del Golfo de Tailandia ha sufrido más cambios que la costa occidental de la Península. Los episodios de erosion han sido recurrentes en areas sujetas a ambientes energéticos, observados, respectivamente, en un 29% y un 11% de la línea de costa occidental y oriental. La pérdida de area costera se estima en 116 ha por año (respectivamente, 91 y 25 ha para las costas occidental y oriental. Lo factores responsables de la erosion son la pérdida de manglares, el aumento en el cultivo de camarón y la disminución del aporte de sedimentos debido a la ubicación de presas en los ríos y la exposición a los vientos del monzón. La erosion fue más moderada en areas con manglar, e incluso se observó progression de manglares en segmentos de costa dominados por manglares. La tasa de progression de manglares se estimó en 37 ha anuales en el este y 5 ha anuales en el oeste. Por tanto, la tasa de erosion neta acumulada en las tres últimas décadas en la costa Sur de Tailandia se estima en 74 ha anuales. Estos resultados también demuestran que la presencia y progression de manglares junto con una tasa de aporte de sedimentos suficiente contribuyen a la estabilidad de las costas.

Una simulación de 33 aós del modelo demográfico mostró que la herbivoría y la turbulencia son los factores más importantes que afectan el éxito de la colonización en una bahía dominada por manglares, mientras que la salinidad y la sedimentación gradual tuvo pocos efectos. *Avicennia* era el género con más éxito en la colonización de llanuras intermareales seguido de *Sonneratia* mientras que *Rhizophora* mostró un exito limitado. Sin embargo, el éxito de colonización variaba con las condiciones ambientales. Por ejemplo, un aumento del nivel del mar reduce el éxito de la colonzación para todos los géneros y la disminución del aporte de agua dulce por el desarrollo de embalses podría incrementar el éxito de colonización de *Avicennia* pero afectar negativamente el éxito colonizador de *Sonneratia*.

En general, este estudio sugiere que el éxito de recolonización demanglares depende tanto de factores ecológicos como hidrológicos. La herbivoría sobre las plántulas y la turbulencia se mostraron como factores capaces de afectar el éxito de la colonización. También, los tres géneros estudiados mostraron grandes diferencias en su capacidad de colonización. La sedimentación gradual mostró pocos efectos sobre la colonización, pero el aumento de la colonización genera habitat para la colonización, particularmente en en segmentos costeros protegidos. La disminución del aporte de sedimentos junto con la conversion de manglares y bosques de playas en estanques de acuacultura aumentan la erosion de segmentos de costa vulnerables. Por tanto, los gestures deben mostrar particular atención al equilibrio entre demandas ecológicas y socioeconómicos para el desarrollo sostenible de la zona costera.

บทสรุป

พลวัตของป่าชายเลนและตะกอนตามแนวพื้นที่ชายฝั่งภาคใต้ของไทย

ป่าชายเลนเป็นสังคมพืชไม่ผลัดใบพบขึ้นอยู่ทั่วไปตามแนวชายฝั่งของประเทศในเขตโซนร้อนและกึ่ง
โซนร้อน โดยเฉพาะอย่างยิ่งบริเวณดินดอนสามเหลี่ยมปากแม่น้ำและอ่าว ซึ่งเป็นจุดที่ตะกอนที่ไหลมา
กับแม่น้ำลำคลองถูกปล่อยลงสู่ทะเล ป่าชายเลนนั้นมีความสำคัญต่อประชาชนที่อาศัยอยู่ตามแนวชายฝั่ง
เป็นอย่างมาก ทั้งทางด้านระบบนิเวศน์ และเศรษฐกิจ-สังคม อาทิ เป็นแหล่งวางไข่และที่อยู่อาศัยของ
สัตว์น้ำนานาชนิดอันเป็นแหล่งอาหารโปรตีนที่สำคัญ ไม้จากป่าชายเลนสามารถนำมาประกอบในการ
สร้างบ้านเรือนและอุปกรณ์ประมง รวมทั้งใช้เป็นฟืนและถ่านในการหุงต้มอาหาร ระบบรากหายใจที่
สลับซับซ้อนช่วยในการดักตะกอน ทำให้เกิดหาดเลนงอกใหม่และช่วยยึดแนวตลิ่ง นอกจากนี้แนวป่า
ชายเลนยังทำหน้าที่เป็นเกราะป้องกันชายฝั่งช่วยกำบังลมพายุและคลื่นขนาดใหญ่เช่น สึนามิ ดังที่เกิดขึ้น
เมื่อวันที่ 26 ธันวาคม 2547 ซึ่งป่าชายเลนช่วยบรรเทาความเสียหายที่เกิดขึ้นต่อชีวิตและทรัพย์สินของ
ชุมชนชายฝั่งได้เป็นอย่างดี

ปัจจุบันสถานะภาพของป่าชายเลนอยู่ในขั้นวิกฤต โดยในภูมิภาคเอเชียตะวันออกเฉียงใต้พื้นที่ป่าชายเลน
ได้ลดลงไปมากกว่าร้อยละ 50 ในระยะเวลา 35 ปีที่ผ่านมา สาเหตุสำคัญที่ทำให้พื้นที่ป่าชายเลนลดลง
ได้แก่ การบุกรุกพื้นที่เพื่อสร้างที่อยู่อาศัยเนื่องจากการขยายตัวของจำนวนประชากร การขุดบ่อเพื่อ
เพาะเลี้ยงสัตว์น้ำ การพังทลายของชายฝั่ง การขาดจิตสำนึก นโยบายรัฐที่คลุมเครือและการขาดการบังคับ
ใช้กฎหมายอย่างจริงจัง จากภาพถ่ายดาวเทียมของประเทศไทยพบว่าตลอดแนวชายฝั่งอ่าวไทยมีนากุ้ง
ตั้งอยู่เรียงรายและมีจำนวนเพิ่มขึ้นทุกปีเพราะไม่มีการจำกัดขอบเขตโดยหน่วยงานใด ซึ่งนากุ้งส่วนใหญ่
จะอยู่ในพื้นที่ป่าชายเลน การลดลงอย่างรวดเร็วของพื้นที่ป่าชายเลนนี้ทำให้ภาครัฐและประชาชนเกิด
ความตระหนักจึงร่วมกันดำเนินโครงการปลูกป่าชายเลนขึ้นในหลายพื้นที่ การปลูกป่าชายเลนในพื้นที่
ป่าเสื่อมโทรมนั้นพบว่าไม่ค่อยประสบผลสำเร็จเพราะสภาพของดินเปลี่ยนแปลงไปมากทำให้ยากต่อการ
ฟื้นฟู และบางแห่งมีความขัดแย้งในสิทธิการถือครองที่ดิน เพื่อหลีกเลี่ยงปัญหาดังกล่าวโครงการปลูกป่า
ชายเลนในระยะหลังจึงหันมาปลูกในพื้นที่ดินเลนงอกใหม่แทน ความสำเร็จของโครงการเหล่านี้จะมีมาก
หรือน้อยแตกต่างกันไปขึ้นอยู่กับความพร้อมและการเตรียมการ มีหลายโครงการที่ลงมือปลูกทันทีโดย
ไม่ได้พิจารณาถึงชนิดพันธุ์ไม้และวิธีการปลูกที่เหมาะสมกับพื้นที่ รวมทั้งละเลยความสามารถในการ
แพร่พันธุ์ตามธรรมชาติของลูกไม้จากผืนป่าที่อยู่ใกล้เคียง

วิทยานิพนธ์ฉบับนี้มีวัตถุประสงค์เพื่อศึกษาการแพร่ขยายเข้าครอบครองหาดเลนงอกใหม่ของลูกไม้จาก
ป่าชายเลนที่อยู่ใกล้หาดเลนงอกใหม่และปัจจัยที่มีผลกระทบต่อการแพร่ขยายดังกล่าว รวมทั้งศึกษาถึง

สาเหตุของปัญหาการกัดเซาะชายฝั่ง โดยหวังว่าข้อมูลที่ได้จะเป็นประโยชน์ต่อการจัดการป่าชายเลนและการจัดการพื้นที่ชายฝั่ง ในขั้นแรกได้ทำการทดลองเพื่อศึกษาถึงผลกระทบของตะกอนและความรุนแรงของกระแสน้ำต่อการรอดตายและการเจริญเติบโตของไม้ชายเลน 3 ชนิดพันธุ์ คือ โกงกาง แสม และลำพู จากนั้นทำการประเมินการเปลี่ยนแปลงพื้นที่ชายฝั่งในภาคใต้ของประเทศไทยระหว่างปี พ.ศ.2504-2543 โดยสังเคราะห์จากข้อมูลการสำรวจแนวชายฝั่งของกรมทรัพยากรธรณีประกอบการศึกษาในภาคสนาม และในขั้นตอนสุดท้ายได้นำข้อมูลต่างๆ มาสร้างแบบจำลองเพื่อเลียนแบบวงจรชีวิตและการแพร่ขยายเข้าครอบครองหาดเลนงอกใหม่ของลูกไม้ชายเลนทั้งสามชนิด

จากการทดลองพบว่าหลังจากที่ลูกไม้สามารถยึดเกาะและงอกบนพื้นเลนได้แล้ว ต้นกล้าที่ได้อาจตายจากการถูกทับถมอย่างฉับพลันของตะกอน และอัตราการตายจะเพิ่มมากขึ้นตามระดับความสูงของตะกอน เมื่อถูกถมด้วยตะกอนสูง 32 ซม. ต้นกล้าของแสมจะตายหมด ส่วนโกงกางและลำพูมีอัตราการรอดตายร้อยละ 30 และ 70 ตามลำดับ ซึ่งแสดงว่าต้นกล้าของแสมมีความอ่อนไหวมากต่อการถูกตะกอนทับถม ในขณะที่ลำพูมีความอ่อนไหวน้อยและยังมีอัตราการเจริญเติบโตที่รวดเร็วกว่าด้วย นอกจากนี้ต้นกล้าอาจตายเพราะผลกระทบจากความรุนแรงของกระแสน้ำได้เช่นกัน ในแปลงปลูกที่มีความหนาแน่นของต้นไม้ที่มีอยู่เดิมน้อยพบว่าต้นกล้าของโกงกางมีอัตราการรอดตายต่ำ แต่ต้นกล้าของแสมและลำพูมีอัตราการรอดตายสูง ลักษณะการรอดตายดังกล่าวสอดคล้องกับพฤติกรรมการเป็นพันธุ์ไม้บุกเบิกของแสมและลำพู และการเป็นพันธุ์ไม้ทดแทนของโกงกาง ผลการทดลองเหล่านี้แสดงให้เห็นถึงความสามารถในการแพร่พันธุ์เข้าครอบครองหาดเลนงอกใหม่ของแสมและลำพู รวมถึงความเหมาะสมของลำพูสำหรับใช้ปลูกในพื้นที่ที่เสี่ยงต่อการตกทับถมของตะกอนสูง

ผลการประเมินการเปลี่ยนแปลงแนวชายฝั่งทะเลของภาคใต้พบว่าในระยะ 30 ปีที่ผ่านมา ฝั่งทะเลด้านอ่าวไทยมีการเปลี่ยนแปลงมากกว่าด้านทะเลอันดามัน โดยฝั่งอ่าวไทยมีการกัดเซาะชายฝั่งเกิดขึ้นเป็นช่วงๆ ระยะทางที่ถูกกัดเซาะคิดเป็นร้อยละ 29 ของความยาวชายฝั่ง ฝั่งทะเลอันดามันมีการกัดเซาะชายฝั่งเช่นกัน มีระยะทางที่ถูกกัดเซาะร้อยละ 11 ของความยาวชายฝั่ง อัตราการกัดเซาะพื้นที่ทั้งหมดคิดเป็นประมาณ 725 ไร่ต่อปี ฝั่งอ่าวไทยถูกกัดเซาะปีละ 569 ไร่ และฝั่งอันดามันถูกกัดเซาะปีละ 156 ไร่ จากการวิเคราะห์ข้อมูลพบว่าสาเหตุสำคัญที่ทำให้เกิดการเซาะชายฝั่งได้แก่ การลดลงของพื้นที่ป่าชายเลน การเพิ่มจำนวนพื้นที่ของนากุ้ง การลดลงของตะกอนที่ไหลลงสู่ทะเลเพราะมีการสร้างเขื่อนกั้นลำน้ำ และลักษณะการเปิดโล่งของพื้นที่ต่ออิทธิพลจากลมมรสุม ทั้งนี้ยังพบว่าการกัดเซาะชายฝั่งเกิดขึ้นน้อยในบริเวณชายฝั่งที่มีป่าชายเลนขึ้นปกคลุม นอกจากนี้ป่าชายเลนในชายฝั่งที่มีคลื่นรบกวนน้อยยังสามารถขยายพื้นที่ได้ด้วยตนเอง โดยฝั่งอ่าวไทยมีป่าชายเลนขยายพื้นที่ด้วยตนเองปีละ 231 ไร่และฝั่งอันดามันประมาณ 31 ไร่ ดังนั้นสถานภาพโดยรวมแล้วในช่วงปี พ.ศ.2504-2543 นั้น ประเทศไทยสูญเสียพื้นที่ไป

ปีละประมาณ 463 ไร่ ผลการศึกษาที่ได้ ชี้ให้เห็นว่าการมีป่าชายเลนตามแนวชายฝั่งและการมีตะกอน
ไหลออกสู่ทะเลอย่างเพียงพอนั้นมีความสำคัญอย่างมากต่อเสถียรภาพของชายฝั่งทะเล

การทดสอบแบบจำลองเพื่อเลียนแบบวงจรชีวิตและการแพร่ขยายเข้าครอบครองหาดเลนงอกใหม่ของ
ลูกไม้ชายเลนในช่วงเวลา 30 ปี พบว่าปัจจัยสำคัญที่มีอิทธิพลต่อความสำเร็จของการแพร่ขยายของลูกไม้
บนหาดเลนงอกใหม่ คือ การถูกกัดกินโดยปู หนอนหรือสัตว์บางชนิด และผลกระทบจากความรุนแรง
ของกระแสน้ำ ส่วนการถูกทับถมด้วยตะกอนทีละเล็กทีละน้อยและความเค็มของน้ำนั้นมีอิทธิพลเพียง
เล็กน้อย ทั้งนี้แสมเป็นพันธุ์ไม้ที่ประสบความสำเร็จในการแพร่ขยายพันธุ์บนหาดเลนงอกใหม่มากที่สุด
ตามด้วยลำพู ส่วนโกงกางนั้นพบว่าแพร่ขยายได้เพียงเล็กน้อย อย่างไรก็ตามความสามารถในการแพร่
ขยายนี้อาจแปรเปลี่ยนไปตามการเปลี่ยนแปลงของสภาวะแวดล้อม โดยผลจากการจำลองสถานการณ์
พบว่าความสำเร็จในการแพร่ขยายของพันธุ์ไม้ทั้ง 3 ชนิดจะลดลงถ้าระดับน้ำทะเลสูงขึ้น และแสมจะ
แพร่ขยายได้มากขึ้นหากกระแสน้ำจืดที่ไหลลงสู่พื้นที่มีปริมาณลดน้อยลงหรือมีการสร้างเขื่อนกั้นแม่น้ำ
แต่เหตุการณ์นี้จะมีผลกระทบอย่างมากต่อลำพูซึ่งเป็นพันธุ์ไม้ที่ไม่ทนต่อสภาวะน้ำเค็มจัด จึงส่งผลให้
ความสามารถในการขยายพื้นที่ลดลงไปกว่าร้อยละ 80

โดยรวมแล้วการศึกษานี้แสดงให้เห็นว่าความสำเร็จในการขยายพื้นที่ด้วยตัวเองของไม้ชายเลนบนหาดเลน
งอกใหม่นั้นขึ้นอยู่กับปัจจัยทั้งทางด้านนิเวศวิทยาและอุทกวิทยา ซึ่งพันธุ์ไม้ชายเลนทั้ง 3 ชนิดแสดง
ความสามารถในการเข้ายึดครองพื้นที่หาดเลนแตกต่างกันไป การกัดกินลูกไม้ของปูหรือสัตว์ชนิดอื่น
และความแรงของกระแสน้ำเป็นปัจจัยที่มีอิทธิพลมากต่อการเข้ายึดเกาะและการงอกบนหาดเลน ในขณะที่
การตกทับถมของตะกอนทีละเล็กทีละน้อยมีผลกระทบต่ออัตราการรอดตายไม่มากนัก แต่ปริมาณของ
ตะกอนที่ตกทับถมอย่างต่อเนื่องเป็นการสร้างหาดเลนว่างให้ป่าชายเลนขยายพื้นที่ออกไป โดยเฉพาะ
บริเวณชายฝั่งที่สงบเงียบ การลดลงของปริมาณตะกอนที่ปล่อยลงสู่ชายฝั่งเพราะมีการสร้างเขื่อน ร่วมกับ
การเปลี่ยนพื้นที่จากป่าชายเลนหรือป่าชายหาดไปเป็นนากุ้ง จะช่วยเร่งให้เกิดการกัดเซาะชายฝั่งมากขึ้น
ดังนั้นผู้บริหารและหน่วยงานที่เกี่ยวข้องควรให้ความสำคัญในสมดุลระหว่างความต้องการใช้ทรัพยากร
เพื่อการพัฒนาเศรษฐกิจและสังคมกับระบบนิเวศให้มากยิ่งขึ้น เพื่อให้การพัฒนาพื้นที่ชายฝั่งมีความยั่งยืน

About the Author

Udomluck Thampanya, was born on 19 August 1962 in Chiang Mai, Northern Thailand. She studied statistics at the Faculty of Science, Chiang Mai University and graduated in 1985. In the same year, she joined the Royal Thai Navy as an officer at the War-game Simulation Division, Institute of Advanced Naval Studies, the Royal Thai Navy, based in Bangkok. After a few years of working, she took a leave to further her study in a field of computer science at the Faculty of Computer Engineering, Chulalongkorn University in Bangkok where she obtained her M.Sc. in 1993.

After graduation, she continued working at the Royal Thai Navy until 1995 before moving to work as a researcher in coastal management at the Coastal Resources Institute, Prince of Songkla University, Hat Yai, Southern Thailand. During 1995-1998, she joined the EU-funded project "Coastal Ecosystems Response to Deforestation-derived Siltation in Southeast Asia (CERDS)" to investigate the impacts of sedimentation on coastal ecosystems. In this project, she was involved mostly in the mangrove research component. She was granted a fellowship from the Netherlands Foundation for the Advancement of Tropical Research (WOTRO) to pursue PhD degree in a sandwich scheme in Environmental Science and Technology (IHE-WAU 1998). Her PhD dissertation integrated several studies on the impacts of sedimentation, hydrodynamics and related factors on mangroves and coastal dynamics.

PUBLICATIONS

Duarte, C.M., Geertz-Hansen, O., Thampanya, U., Terrados, J., Fortes, M.D., Kmap-Neilsen, L., Borum, J., Boromthanarat, S. 1998. Relationship between sediment conditions and mangrove *Rhizophora apiculata* seedling growth and nutrient status. Marine Ecology Progress Series 175: 277-283.

Duarte, C.M., Thampanya, U., Terrados, J., Geertz-Hansen, O. and Fortes, M.D. 1999. The determination of the age and growth of SE Asian mangrove seedlings from internodal counts. Mangrove and Salt Marsh 3:251-257.

Panapitukkul, N., Duarte, C.M., Thampanya, U., Terrados, J., Keowvongsri, P., Geertz-Hansen, O., Srichai, N. and Boromthanarat, S. 1998. Mangrove colonization: Mangrove progression over the growing Pak Phanang (SE Thailand) mud flat. Estuarine, Coastal and Shelf Science 47: 51-61.

Terrados, J., Duarte, C.M., Fortes, M.D., Borum, J., Agawin, N. S.R., Bach, S., Thampanya, U., Kmap-Neilsen, L. 1998. Changes in community structure and biomass of seagrass communities along gradients of siltation in SE Asia. Estuarine, Coastal and Shelf Science 46: 757-768.

Terrados, J., Thampanya, U., Srichai, N., Keowvongsri, P., Geertz-Hansen, O., Boromthanarat, S., Panapitukkul, N. and Duarte, C.M. 1997. The effect of increased sediment accretion on the survival and growth of *Rhizophora apiculata* seedlings. Estuarine, Coastal and Self Science 45: 697-701.

Thampanya U., Vermaat, J.E. and Terrados, J. 2002. The effect of increasing sediment accretion on the seedlings of three common Thai mangrove species. Aquatic Botany 74: 315-325.

Thampanya, U., Vermaat, J.E. and Duarte, C.M. 2002. Colonization success of common Thai mangrove species as a function of shelter from water movement. Marine Ecology Progress Series 237: 111-120.

Thampanya U., Vermaat, J.E. 2001. Impact of sedimentation on mangrove dynamics along Thai coastlines. European Tropical Forest Research Network No.33: Forest and Water, Spring-Summer 2001, pp17-18

Printed and bound by CPI Group (UK) Ltd, Croydon, CR0 4YY

22/10/2024

01777637-0017